AF581701

Analysis of the Harmonic Performance of Power Converters and Electrical Drives

Analysis of the Harmonic Performance of Power Converters and Electrical Drives

Editors

Vito Giuseppe Monopoli
Abraham Marquez

MDPI • Basel • Beijing • Wuhan • Barcelona • Belgrade • Manchester • Tokyo • Cluj • Tianjin

Editors

Vito Giuseppe Monopoli
Departement of Electrical and Information Engineering
Politecnico di Bari
Bari
Italy

Abraham Marquez
Ingenieria Electrónica
Universidad de Sevilla
Sevilla
Spain

Editorial Office
MDPI
St. Alban-Anlage 66
4052 Basel, Switzerland

This is a reprint of articles from the Special Issue published online in the open access journal *Energies* (ISSN 1996-1073) (available at: www.mdpi.com/journal/energies/special_issues/analysis_harmonic_power_converters_electrical_drives).

For citation purposes, cite each article independently as indicated on the article page online and as indicated below:

LastName, A.A.; LastName, B.B.; LastName, C.C. Article Title. *Journal Name* **Year**, *Volume Number*, Page Range.

ISBN 978-3-0365-1502-1 (Hbk)
ISBN 978-3-0365-1501-4 (PDF)

Contents

About the Editors

Vito Giuseppe Monopoli

Vito Giuseppe Monopoli (Senior Member, IEEE) received M.Sc. and Ph.D. degrees in Electrical Engineering from the Politecnico di Bari, Bari, Italy, in 2000 and 2004, respectively. He is currently an Assistant Professor with the Politecnico di Bari. His research interests include multilevel converters and the analysis of harmonic distortion produced by power converters and electrical drives. He is particularly interested in innovative control techniques for power converters. He is also a member of the IEEE Industry Applications Society, the IEEE Industrial Electronics Society, and the IEEE Power Electronics Society.

Abraham Marquez

Abraham Marquez (amarquez@ieee.org) earned his Ph.D. degree in Telecommunications Engineering from Universidad de Sevilla, Spain, in 2019.

His research interests include modulation techniques, multilevel converters, the model-based predictive control of power converters and drives, renewable energy sources, and extending the lifetime of power devices. As a coauthor, we was a recipient of the 2015 Best Paper Award of IEEE Industrial Electronics Magazine. He is a member of the IEEE Industrial Electronics Society, and Member of IEEE.

Preface to "Analysis of the Harmonic Performance of Power Converters and Electrical Drives"

Power converters have progressively become the most efficient and attractive solution in recent decades in many industrial sectors, ranging from electric mobility, aerospace applications to attain better electric aircraft concepts, vast renewable energy resource integration in the transmission and distribution grid, the design of smart and efficient energy management systems, the usage of energy storage systems, and the achievement of smart grid paradigm development, among others.

In order to achieve efficient solutions in this wide energy scenario, over the past few decades, considerable attention has been paid by the academia and industry in order to develop new methods to achieve power systems with maximum harmonic performance aiming for two main targets. Analysis of the harmonic performance of power converters and electrical drives has become a critical issue for contemporary power applications in order to reduce power system costs and power density, as well as meeting national and international standards. These objectives are fundamental in applications where the size and weight of the power system is critical (i.e., aerospace or electric vehicle sectors).

This becomes critical in applications such as aerospace or electric mobility where the power converters are on-board systems. On the other hand, current standards are becoming more and more strict in order to reduce EMI and EMC noise, as well as meeting minimum power quality requirements (i.e., grid code standards for grid-tied power systems).

For all these reasons, the development of advanced modulation and control techniques paying special attention to the harmonic performance of power converters is a key research topic. The purpose of this book is to summarize some interesting recent research results related to this issue. This research will push the academia and industry to speed up the development of new solutions for present and future power applications with strict power quality, size, weight and economical cost requirements. The expected audience of this book are academia and industry members highly specialized in advanced modulation and control methods aiming to reduce the harmonic impact of power system operations.

The co-editors of this book are Dr. Vito G. Monopoli from Politecnico di Bari (Italy) and Dr. Abraham Marquez Alcaide from Universidad de Sevilla (Spain).

Firstly, in the work titled "Common-Mode Voltage Harmonic Reduction in Variable Speed Drives Applying a Variable-Angle Carrier Phase-Displacement PWM Method", a generalized PWM method where the carriers present a variable phase-displacement is developed in order to achieve the CMV reduction without any external component and/or passive filtering technique.

In a second work, entitled "Analysis and Implementation of Multilevel Inverter for Full Electric Aircraft Drives", a study considering a multilevel cascaded H-bridge converter for a permanent magnet synchronous motor drive for application in fully electric aircrafts is summarized; the devices represents a very suitable solution for this application.

On the other hand, in the work titled "Frequency-Domain Modeling of Harmonic Interactions in Voltage-Source Inverters with Closed-Loop Control", the authors develop a complete, nonlinear modeling approach for voltage-source inverters in the frequency domain, taking into account the harmonic components introduced into the system from the inputs and from a nonlinear digital asymmetrical regularly-sampled double-edge PWM (AD-PWM). The objective of the work is to analyze how the harmonics propagate through the nonlinear system in steady states.

In addition, in the work titled "Coordinated Control of Active and Reactive Power Compensation for Voltage Regulation with Enhanced Disturbance Rejection Using Repetitive Vector-Control", the authors address the enhanced coordinated control of active and reactive power injected in a distribution grid for voltage regulation. Voltage drop mitigation is evaluated with power injection based on local features, such loads and disturbances of each connection. In order to ensure disturbance rejections such as harmonic components in the grid voltages, a repetitive vector-control scheme is proposed.

Finally, the work entitled "Equivalent Phase Current Harmonic Elimination in Quadruple Three-Phase Drives Based on Carrier Phase Shift Method"addresses a new method for the modelling of equivalent phase currents in multiple three-phase drives with the double-integral Fourier analysis method. A new carrier-based pulse-width modulation (CPWM) method is introduced to reduce the equivalent phase current harmonics by applying the proper carrier phase angle to each subsystem in the three-phase drives.

The co-editors would like to acknowledge the financial support of Conserjeria de Ecónimia, Conocimiento, Empresa u Universidades of the Junta de Andalucia under the projects P18-RT-1340 and PAIDI –DOCTOR "Contratacion de Personal Investigador Doctor (convocatoria 2019) 43 Contratos Capital Humano Linea 2, Paidi 2020.

Vito Giuseppe Monopoli, Abraham Marquez

Editors

icle

ommon-Mode Voltage Harmonic Reduction in Variable Speed rives Applying a Variable-Angle Carrier Phase-Displacement WM Method

raham Marquez Alcaide [1,*,†], Vito Giuseppe Monopoli [2,†], Xuchen Wang [3,†], Jose I. Leon [1,4,†], ampaolo Buticchi [3,†], Sergio Vazquez [1,†], Marco Liserre [5,†] and Leopoldo G. Franquelo [1,4,†]

1 Electronic Engineering Department, Universidad de Sevilla, 41004 Sevilla, Spain; jileon@us.es (J.I.L.); sergi@us.es (S.V.); lgfranquelo@us.es (L.G.F.)
2 Department of Electrical and Information Engineering, Politecnico di Bari, 70126 Bari, Italy; vitogiuseppe.monopoli@poliba.it
3 Zhejiang Key Laboratory on the More Electric Aircraft Technologies, University of Nottingham Ningbo China, Ningbo 315100, China; xuchen.wang@nottingham.edu.cn (X.W.); Giampaolo.Buticchi@nottingham.edu.cn (G.B.)
4 Department of Control Science and Engineering, Harbin Institute of Technology, Harbin 150001, China
5 Power Electronics Department, Christian-Albrechts Universität zu Kiel, 24118 Kiel, Germany; ml@tf.uni-kiel.de
* Correspondence: amarquez@ieee.org
† These authors contributed equally to this work.

tion: Marquez Alcaide, A.; nopoli, V.G.; Wang, X.; Leon, J.I.; cchi, G.; Vazquez, S.; Liserre, M.; nquelo, L.G. Common-Mode age Harmonic Reduction in able Speed Drives Applying a able-Angle Carrier Phase-placement PWM Method. *Energies* 1, *14*, 2929. https://doi.org/390/en14102929

demic Editor: Mario Marchesoni

eived: 6 April 2021
epted: 5 May 2021
lished: 19 May 2021

lisher's Note: MDPI stays neutral n regard to jurisdictional claims in lished maps and institutional affil-ons.

Abstract: Electric variable speed drives (VSD) have been replacing mechanic and hydraulic systems in many sectors of industry and transportation because of their better performance and reduced cost. However, the electric systems still face the issue of being considered less reliable than the mechanical ones. For this reason, researchers have been actively investigating effective ways to increase the reliability of such systems. This paper is focused on the analysis of the common-mode voltage (CMV) generated by the operation of the VSDs which directly affects to the lifetime and reliability of the complete system. The method is based on the mathematical description of the harmonic spectrum of the CMV depending on the PWM method implementation. A generalized PWM method where the carriers present a variable phase-displacement is developed. As a result of the presented analysis, the CMV reduction is achieved by applying the PWM method with optimal carrier phase-displacement angles without any external component and/or passive filtering technique. The optimal values of the carrier phase-displacement angles are obtained considering the minimization of the CMV total harmonic distortion. The resulting method is easily implementable on mostly off-the-shelf mid-range micro-controller control platforms. The strategy has been evaluated in a scaled-down experimental setup proving its good performance.

Keywords: power converters; harmonic distortion; pulse-width modulation; metaheuristic search algorithms

1. Introduction

Power converters are widely used in all energy scenarios such as motor drives, renewable energy sources integration, power quality applications, energy storage systems and efficient electrical transportation. From low- to high-power applications, grid-connected and stand-alone systems can be found [1]. As a general trend, the applications are moving forward from mechanical and/or hydraulic systems to electrical drives because of the increase in performance, pricing cutting, size and weight reduction and maintenance tasks simplification [2]. As a consequence, the variable-speed drive (VSD) actually plays one of the most important roles in the power electronics field.

Power electronics have gone through a huge development in the last few decades, especially in recent years. New power devices technologies, efficient power converter topologies as well as high-performance control and modulation strategies have emerged a result of joint research between academia and industry [3–5]. As an example, multip advanced multi-level topologies have been demonstrated as a good candidate for t motor drive application [6]. However, despite most of these new technologies being fu available, the industrial impact of these solutions by the industry is still limited. Eith due to strict design requirements, limitations in system implementation, the requir power and voltage levels or the slow implementation pace in the industry, the conventio three-phase two-level IGBT-based power converter is the most widely used topology most of the industrial products. In Figure 1, the traditional scheme of a motor drive represented including a three-phase two-level IGBT-based power converter. In Figure the possible capacitive coupling to ground are also represented, which helps to develop circuit to model the electrical behavior of the system.

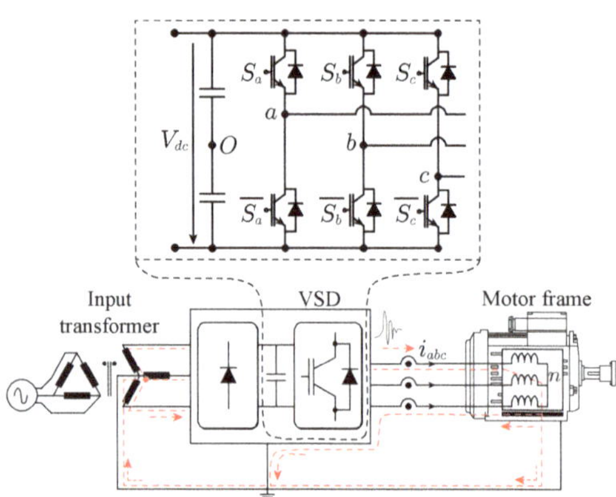

Figure 1. Traditional three-phase two-level converter for a motor drive application.

The control and modulation techniques for VSDs have been well-known by indust and academia for decades and multiple options can be found in the literature [4]. T use of a power converter as VSD presents many advantages such as enhanced dynam performance, high efficiency due to the accurate control of flux and current and improv system efficiency at partial loads [7]. From the modulation point of view, among t possible alternatives, the mainstream solution to operate the two-level VSD is to apply pulse-width modulation (PWM) technique based on a bipolar PWM strategy per pha using only one high-frequency triangular carrier [3]. This PWM implementation is t simplest one being a straight-forward modulation method because most of availab control platforms include dedicated hardware peripherals to deploy the PWM techniq to generate the switching signals of the power semiconductors [8]. In Figure 2a, th traditional PWM method for a three-phase two-level power converter is shown.

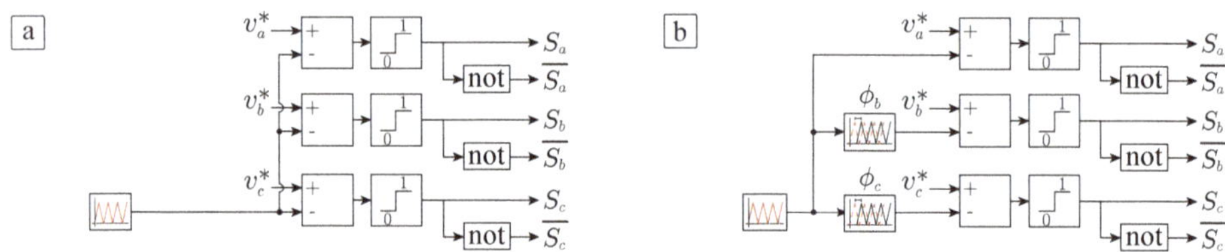

Figure 2. Modulation method of a three-phase two-level motor drive (**a**) Traditional PWM method with a single carrier (**b**) PWM technique with three carriers.

Despite the advantages of power converters based on VSD, several drawbacks asso ated with the high-switching frequency affects to the performance and reliability of t system [9–11]. For instance, a conducted and radiated electromagnetic interference (EMI) is introduced. Additionally, the shaft voltage and bearing currents phenomenon for the electrical machines appear [12–16]. On the other hand, the common-mode voltage (CMV) generated by the VSD directly affects the reliability of the motor because of the degradation of the bearings. It has been reported that CMV is the main cause of the bearing degradation which represents more than 50% of the motor failures [9,10,17,18]. In order to face this issue, different mathematical approaches have been developed in order to estimate and to predict the early failure of these motor components [19–22]

Multiple solutions can be found in the literature regarding how to eliminate or mitigate the negative effects provoked by the inherent CMV created by the VSD operation to drive the machines. For instance, the introduction of the passive filters as well as active circuits for the CMV elimination have been explored by the academia [16,23–28]. In case of using passive elements filters, especially for high-power drives, the size and cost is increased. On the other hand, considering active common-noise canceler topologies, although good results are achieved the introduction of extra passive elements as well as power devices (and their auxiliary systems and drivers) are required. Therefore, both solutions increase the cost of the whole system, its complexity, volume and weight. On the other hand, the CMV mitigation by developing new modulation techniques has been also explored by the academia and industry. For instance, in [17,25,27,29–32] the proper switching patter selection based on space-vector modulation methods is proposed. Taking into account carrier-based PWM methods, as shown in Figure 2b, a PWM technique with three triangular carriers with phase-displacements equal to $[\phi_a, \phi_b, \phi_c] = [0, 120, 240]^\circ$ was proposed in [33], where these carrier phase-displacement angles are found after an empirical trial and error method. In [34], an adaptive sampling time tri-carrier PWM was also proposed, where an effective CMV reduction is achieved for higher modulation indexes.

This paper provides a generalized analysis of the CMV harmonic spectrum generated during the operation of the VSD considering a PWM method with any carrier phase-displacement angles defined by $[\phi_a, \phi_b, \phi_c]$. In this work, the carrier phase-displacement angles are chosen considering the minimization of the whole CMV harmonic content. As a consequence of the analysis, the CMV harmonic distortion mitigation without using any external component and/or passive filtering technique is achieved. The proposed method is easily implementable on mostly off-the-shelf mid-range micro-controller control platforms.

2. Effect of the Common-Mode Voltage on the Motor Bearing Degradation

Figure 3a shows a section of a permanent-magnet synchronous machine (PMSM) highlighting the most important mechanical parts. Considering the bearing current effect caused by common-mode voltage (CMV) at neutral point of machine terminals (n), the machine can be modeled with a group of parasitic capacitors as shown in Figure 3b. The rotor shaft is capacitively coupled with the machine stator windings and stator frame, respectively. These capacitances act as a voltage divider, which leads to a rate of CMV on the rotor shaft [22]. While the shaft voltage exceeds the threshold voltage of greasing film (normally between 1.5 V and 30 V), the high dv/dt on the shaft will generate a high current pulse flowing from the bearing to the ground, which is regarded as the electric discharge machining (EDM) [11,20]. It has negative influence on the drive system with the resulting electromagnetic interference (EMI) and grounding failure. In addition, the bearing might be damaged by the circulating currents in the machine. Proper PWM techniques can be adopted to reduce the common-mode variations, which can avoid bearing failures and reduce the EMI [35].

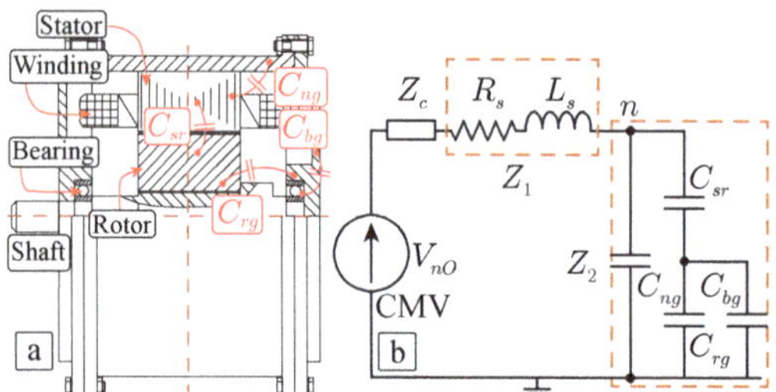

Figure 3. (**a**) PMSM machine section. (**b**) Equivalent impedance model of a PMSM machine.

As mentioned, in Figure 3b the common-mode equivalent circuit of the PMSM driv system is represented. In the figure, Z_c is the common-mode impedance of the cable, an R_s and L_s are the stator winding resistance and stator winding inductance, respectively. T simplify the model, Z_1 is the equivalent common-mode impedance of R_s and L_s. Z_2 is th equivalent common-mode impedance of the stator-to-earth parasitic capacitance, whic consists of the parasitic capacitance between stator and earth C_{ng}, the parasitic capacitanc between stator and rotor C_{sr}, the parasitic capacitance betwe and earth C_{rg} and the parasit capacitance of bearing C_{bg}.

The CMV of stator neutral point n can be calculated as

$$V_{ng} = \frac{Z_2}{Z_c + Z_1 + Z_2} CMV$$

In the PMSM, the shaft voltage V_b is expressed as

$$V_b = \frac{C_{sr}}{C_{sr} + C_{rg} + C_{bg}} V_{ng}$$

Since the impedance Z_x varies with the frequency, then the above equation can b expressed in the frequency domain as:

$$V_{ng}(\omega) = \frac{Z_2(\omega)}{Z_c(\omega) + Z_1(\omega) + Z_2(\omega)} CMV(\omega) = K \cdot CMV(\omega)$$

Since factor K is constant, if CMV harmonic content is reduced, it is possible t mitigate the collateral effects in the generated V_{ng} and as consequence in V_b. In this wa the minimization of the CMV harmonic distortion represents a challenge to be overcome i order to extend the PMSM lifetime.

3. Common-Mode Voltage Harmonic Description in a Two-Level VSI Operating as a Motor Drive

The CMV of the power converter shown in Figure 1 is defined as the voltage betwee the neutral point in the load (n) and the middle point in the dc-link. It may happen that th neutral point n is not accessible to take the measurement and, therefore, the CMV can no be directly obtained. As an alternative, the CMV can be calculated as the summation of th phase voltages divided by the number of inverter phases as:

$$CMV = V_{n0} = \frac{v_{aO} + v_{bO} + v_{cO}}{3}$$

where v_{xO} is the phase voltage between the phase x terminal and the point O (middle poir of the dc-link).

In order to obtain an analytical expression of the CMV, a double Fourier series expre sion can be used, leading to:

$$x(t) = \frac{A_{00}}{2} + \sum_{n=1}^{\infty}\left[A_{0n}\cos(n\omega_0 t) + B_{0n}\sin(n\omega_0 t)\right]$$
$$+ \sum_{m=1}^{\infty}\sum_{n=1}^{\infty}\left[A_{mn}\cos(m\omega_c + n\omega_0 t) + B_{mn}\sin(m\omega_c + n\omega_0 t)\right] \quad (5)$$

where coefficients A_{00} represents the average value and A_{0n} and B_{0n} represent the base-bands harmonic content. Coefficients A_{mn}, B_{mn} represent the side-bands harmonic content. All these coefficients are calculated using the double Fourier integral [3].

Considering the traditional single-carrier PWM modulation approach shown in Figure 2a, after the Fourier coefficients calculation, the phase voltage in a VSD can be described using the double Fourier series as:

$$v_{xO}(t) = \frac{M_x V_{dc}}{2}\cos(\omega_0 t + \theta_x)$$
$$+ \frac{2V_{dc}}{\pi}\sum_{m=1}^{\infty}\sum_{n=1}^{\infty}\left[\frac{1}{m}J_n\left(\frac{m\pi}{2}M_x\right)\sin\left(\frac{(m+n)\pi}{2}\right)\cos\left(m\left(\omega_c t + \phi_x\right) + n\left(\omega_0 t + \theta_x\right)\right)\right] \quad (6)$$

where M_x and θ_x are, respectively, the modulation index and the phase angle of phase x ($x = a, b, c$). $J_n(z)$ is the first kind Bessel function of z and order n.

In a generalized PWM strategy, a triangular carrier can be considered associated to each converter phase. In this way, as shown in Figure 3b, three carrier signals are considered, where ϕ_x is the phase displacement of the carrier associated to phase x. It is important to notice that only the fundamental frequency as well as side-bands harmonic content appear because the base-band harmonics are canceled.

From (6), it is possible to calculate the magnitude of each side-band harmonic component of the phase voltage v_{xO} as:

$$V_{xO}(m,n) = \frac{2V_{dc}}{m\pi}J_n\left(\frac{m\pi}{2}M_x\right)\sin\left(\frac{(m+n)\pi}{2}\right) \quad (7)$$

From this result, each harmonic component of the CVM in the VSD can be directly evaluated applying (6) using the corresponding components in the phase voltages $V_{xO}(m,n)$. Each harmonic order k can be defined as $k = mR + n$, where $R = \omega_c/\omega_0$. In addition, for the sake of simplicity, it can be considered that the modulation index of all the phases are equal ($M_a = M_b = M_c = M$). As the phase voltages can be described using the equivalent Fourier coefficients and considering that ϕ_a is fixed to 0, the side-band harmonic components of the CMV can described as

$$v_{nO}(m,n) = CMV(m,n) = \frac{V_{dc}}{3m\pi}J_n\left(\frac{m\pi}{2}M\right)\sin\left(\frac{(m+n)\pi}{2}\right)$$
$$\left[\left(1 + e^{j(m\phi_b + n\theta_b)} + e^{j(m\phi_c + n\theta_c)}\right)e^{j(m\omega_c + n\omega_0)t} + \left(1 + e^{-j(m\phi_b + n\theta_b)} + e^{-j(m\phi_c + n\theta_c)}\right)e^{-j(m\omega_c + n\omega_0)t}\right] \quad (8)$$

To illustrate this, a example of the CMV harmonic spectrum is represented in Figure 4 where the magnitude of each harmonic component has been represented following the notation given (8). This example has been illustrated considering that $R = 21$. In this sense, in the first harmonic group ($m = 1$) (and all odd harmonics groups) there are only side-bands with even values of n. Similarly, in the second harmonic group $m = 2$ (and all even harmonic groups) the side-bands present non-zero values with odd values of n.

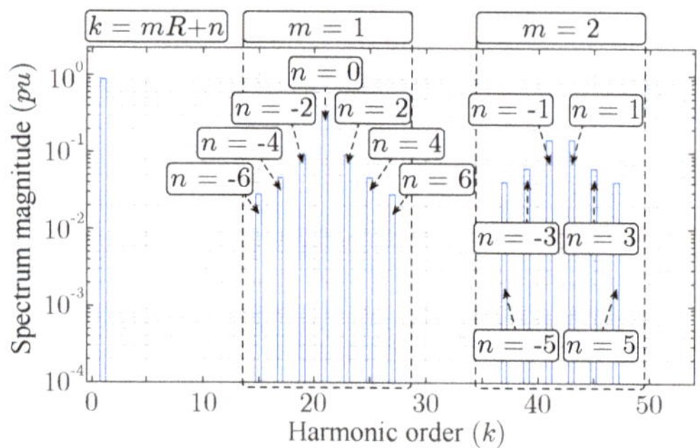

Figure 4. Harmonic spectrum definition of the CMV.

4. Relation between the Generalized Carrier Phase-Displacement PWM Method and the CMV Harmonic Content

From the expressions introduced previously, it is clear that the CMV harmonic conte highly depends on the carrier phase-displacement angles of the PWM method. From (8), a assuming the phase a as reference for convenience (i.e., $\phi_a = 0$, $\theta_a = 0$), the magnitude o generic harmonic component of the $CMV(m,n,\phi_b,\phi_c)$, considering $M_a = M_b = M_c =$ can be re-written as:

$$CMV(m,n,\phi_b,\phi_c) = V_{nO}(m,n,\phi_b,\phi_c) = 2A(m,n)P(m,n,\phi_b,\phi_c)$$

where factors $A(m,n)$ and $P(m,n,\phi_b,\phi_c)$ are defined as:

$$A(m,n) = \frac{V_{dc}}{3m\pi} J_n\left(\frac{m\pi}{2}M\right)\sin\left(\frac{(m+n)\pi}{2}\right) \quad (1$$

$$P(m,n,\phi_b,\phi_c) = \sqrt{3 + 2\cos(m\phi_b + n\theta_b) + 2\cos(m\phi_c + n\theta_c) + 2\cos(m\phi_b + n\theta_b - m\phi_c - n\theta_c)} \quad (1$$

In order to find the values of the carrier phase-displacement angles ϕ_b and ϕ_c th guarantee the minimum magnitude for a specific harmonic component (m,n), it is requir to develop the analytical expressions of the partial derivatives with respect to ϕ_b and ϕ_c. yields to:

$$\frac{\partial V_{nO}(m,n,\phi_b,\phi_c)}{\partial \phi_b} = -\frac{A(m,n)(2m\sin(m\phi_b + n\theta_b - m\phi_c - n\theta_c) + 2m\sin(m\phi_b + n\theta_b))}{\sqrt{3 + 2\cos(m\phi_b + n\theta_b) + 2\cos(m\phi_c + n\theta_c) + 2\cos(m\phi_b + n\theta_b - m\phi_c - n\theta_c)}} \quad (1$$

Then, forcing (12) to be equal to 0:

$$\begin{aligned} \frac{\partial V_{nO}(m,n,\phi_b,\phi_c)}{\partial \phi_b} &= 0 \\ \phi_b &= \frac{1}{m}\left(\frac{m\phi_c}{2} - n\theta_b + \frac{n\theta_c}{2} + k\pi\right) \\ \phi_c &= \frac{2(m\phi_b + n\theta_b) - n\theta_c + 2k\pi}{m} - \frac{\pi + n\theta_c}{m} \end{aligned} \quad (1$$

Analogously, the partial derivative respect to ϕ_c and its solution can be obtained as

$$\frac{\partial V_{nO}(m,n)}{\partial \phi_c} = \frac{A(m,n)(2m\sin(m\phi_b + n\theta_b - m\phi_c - n\theta_c) - 2m\sin(m\phi_c + n\theta_c))}{\sqrt{3 + 2\cos(m\phi_b + n\theta_b) + 2\cos(m\phi_c + n\theta_c) + 2\cos(m\phi_b + n\theta_b - m\phi_c - n\theta_c)}} \quad (1$$

$$\frac{\partial V_{nO}(m,n)}{\partial \phi_c} = 0$$
$$\phi_b = \frac{2(m\phi_c + n\theta_c) - n\theta_b + 2k\pi}{m}$$
$$\phi_c = \frac{1}{m}\left(\frac{m\phi_b}{2} - n\theta_c + \frac{n\theta_b}{2} + k\pi\right) \tag{15}$$

Particularizing both systems (13) and (15) for $m = 1, n = 0$ (because it is the dominant harmonic component in the CMV harmonic spectrum) and $\theta_b = 120°$, $\theta_c = 240°$ which represents the usual case in a three-phase system, it can be seen that both systems of equations achieve two stationary points where $[\phi_b, \phi_c]$ is equal to $[120, 240]°$ or $[240, 120]°$. It is important to notice that this result does not depend on the modulation index M.

This fact is also proven if expression (9) is represented particularized for the harmonic component described by $m = 1$ and $n = 0$. As it is shown in Figure 5a, represented using MatLab by implementing the mathematical description of the harmonic components, the magnitude of this dominant harmonic component is zero in both stationary points ($[\phi_b, \phi_c]$ equal to $[120, 240]°$ and $[240, 120]°$) whereas it is maximum when the traditional PWM technique is used ($\phi_b = \phi_c = 0°$). Additionally, Figure 5b,c represent the values of expression (9) particularized for $m = 1$ and $n = -2$, and $m = 1$ and $n = 2$, respectively. As it is shown in both figures, the carrier phase-displacement angles that achieve the cancellation of the dominant harmonic component do not lead to the cancellation of all other harmonic components. As can be observed in Figure 5, depending on the carrier phase-displacement angles solution that is chosen ($[\phi_b, \phi_c]$ equal to $[120, 240]°$ or $[240, 120]°$), one of the harmonic components with $m = 1$ and $n = \pm 2$ is zero but the other is maximum.

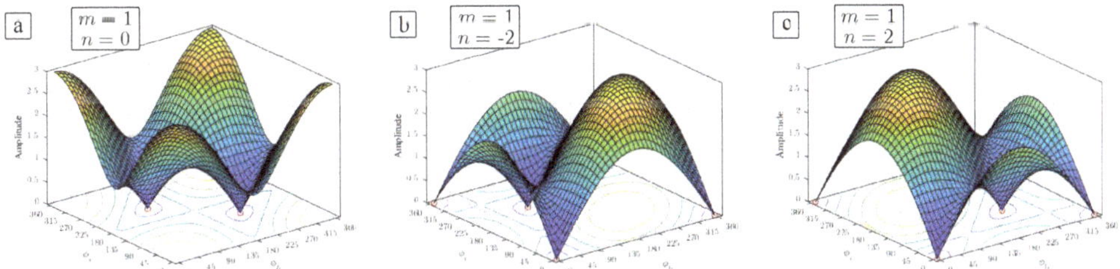

Figure 5. Evaluation of the Equation (9) to evaluate $V_{nO}(m, n, \phi_b, \phi_c)$ considering (**a**) $m = 1, n = 0$ (**b**) $m = 1, n = -2$ (**c**) $m = 1, n = 2$.

As a practical example to show this concept, assuming a modulation index equal to $M = 0.8$ and the traditional carrier phase-displacement angles $[\phi_a, \phi_b, \phi_c] = [0, 0, 0]°$, the magnitude of the harmonic components $V_{nO}(m, n, \phi_b, \phi_c)$ are:

$$\begin{aligned}
V_{nO}(1,0,0,0) &= A(1,0)P(1,0,0,0) && = 0.1363V_{dc} \cdot 3 = 40.89V_{dc}[V] \\
V_{nO}(1,-2,0,0) &= A(1,-2)P(1,-2,0,0) && = 0.0366V_{dc} \cdot 0 = 0[V] \\
V_{nO}(1,2,0,0) &= A(1,2)P(1,2,0,0) && = 0.0366V_{dc} \cdot 0 = 0[V]
\end{aligned} \tag{16}$$

However, if the carrier phase-displacement angles are equal to $[\phi_a, \phi_b, \phi_c] = [0, 120, 240]°$, it leads to:

$$\begin{aligned}
V_{nO}(1,0,120,240) &= A(1,0)P(1,0,120,240) && = 0.1363V_{dc} \cdot 0 = 0[V] \\
V_{nO}(1,-2,120,240) &= A(1,-2)P(1,-2,120,240) && = 0.0366V_{dc} \cdot 0 = 0[V] \\
V_{nO}(1,2,120,240) &= A(1,2)P(1,2,120,240) && = 0.0366V_{dc} \cdot 3 = 0.1098V_{dc}[V]
\end{aligned} \tag{17}$$

Summarizing, although the elimination of the dominant harmonic component ($m =$ $n = 0$) is performed by the PWM method with three carriers applying one of the availab possible carrier phase-displacement solutions, the appearance of a side-band harmon component located at another frequency is unavoidable. However, it is important to noti that this non-zeroed harmonic component presents a lower magnitude than the domina harmonic component located at the frequency described with $m = 1$ and $n = 0$.

Important Remark: It has to be noticed that the use of carrier phase-displaceme angles in the PWM method equal to $[0, 120, 240]^\circ$ is not new. In [33], these angles we applied in a PWM method achieving a reduction in the CMV. However it was not demo strated if this solution is the best one to be chosen in order to achieve maximum CM harmonic performance, which is the main objective of this work. In fact, during the wor it will be demonstrated that the solution proposed in [33] is not optimal in terms of CM harmonic distortion reduction.

5. Proposed Generalized Carrier Phase-Displacement PWM Method to Improve the CMV Total Harmonic Distortion

In order to mitigate the negative CMV effect on the motor bearings, instead of elimina ing one specific harmonic content in the CMV harmonic spectrum as proposed in [33], it more convenient to consider the total harmonic distortion (THD) of the CMV respect to th dc level as figure of merit to be minimized by the application of carrier phase-displaceme angles in the PWM method. In this way, the CMV_{THD} can be defined as:

$$CMV_{THD} = \frac{2}{V_{dc}} \sqrt{\sum_{m=1}^{N-1} \sum_{n=-j}^{j} CMV(m, n, \phi_b, \phi_c)^2} \quad (1$$

It is possible to determine the carrier phase-displacement angles in the PWM metho that minimize the resulting CMV_{THD}. The angles can be obtained with an exhausti search evaluating (18) applying all the possible values of the carrier phase-displaceme angles for different values of the modulation index. This process has been carried o assuming a modulation index range from $M = 0.1$ to $M = 1.15$ and calculating (1 considering the first three harmonic groups ($m \leq 3$) with $|n| \leq 6$.

Figure 6 shows the evaluation of CMV_{THD} with different discrete values of the mo ulation index. As can be observed from the obtained results, the CMV_{THD} generated b the VSD applying the PWM method with $[0, 120, 240]^\circ$ carrier-phase displacement angl is minimum when the modulation index range between 0.3 and 1.15. However, for lo modulation index values, the minimum value of the CMV_{THD} is achieved with oth carrier phase-displacement angles. This result represents an improvement of the metho presented in [33], where the carrier-phase displacement angles equal to $[0, 120, 240]^\circ$ a claimed to be always optimal in terms of the CMV mitigation. As a summary of the o tained results, the corresponding CMV_{THD} values are represented in Figure 7. In the figur three different options are represented, taking into account the carrier phase-displaceme angles in the PWM method. On one hand, the values obtained applying the PWM metho with carrier phase-displacement angles equal to $[0, 0, 0]^\circ$ are higher compared with th other methods. Applying the carrier phase-displacement angles equal to $[0, 120, 240$ achieves good results but the CMV_{THD} is not minimized in the whole modulation rang In order to achieve the minimum value of the CMV_{THD} it is required to apply variabl angle carrier phase-displacement angles in the PWM method. The optimal values of thes angles to achieve the minimization of the CMV_{THD} depending on the modulation inde are represented in Figure 8.

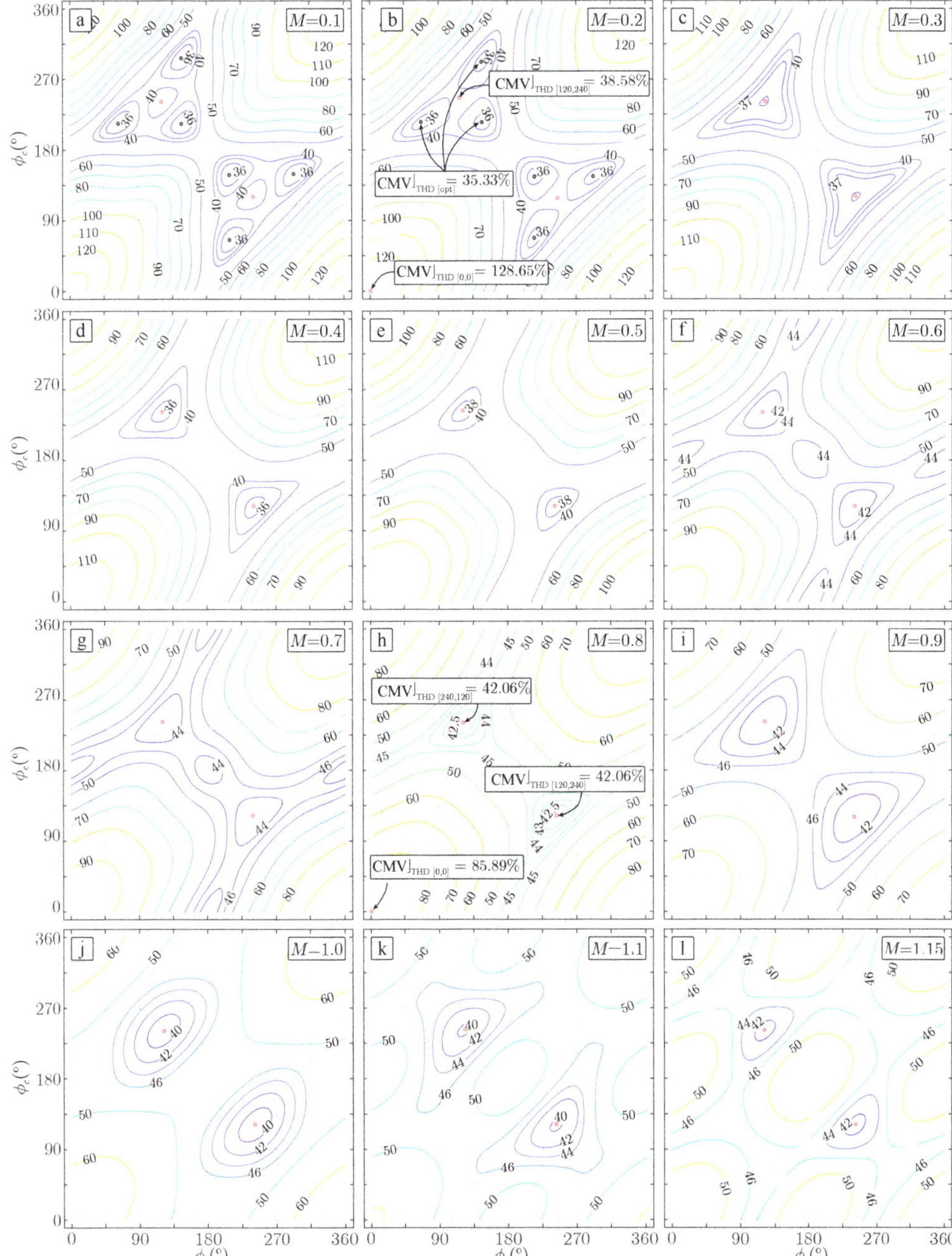

Figure 6. Evaluation of the CMV_{THD} with different values of the modulation index (**a**) M = 0.1 (**b**) M = 0.2 (**c**) M = 0.3 (**d**) M = 0.4 (**e**) M = 0.5 (**f**) M = 0.6 (**g**) M = 0.7 (**h**) M = 0.8 (**i**) M = 0.9 (**j**) M = 1 (**k**) M = 1.1 (**l**) M = 1.15.

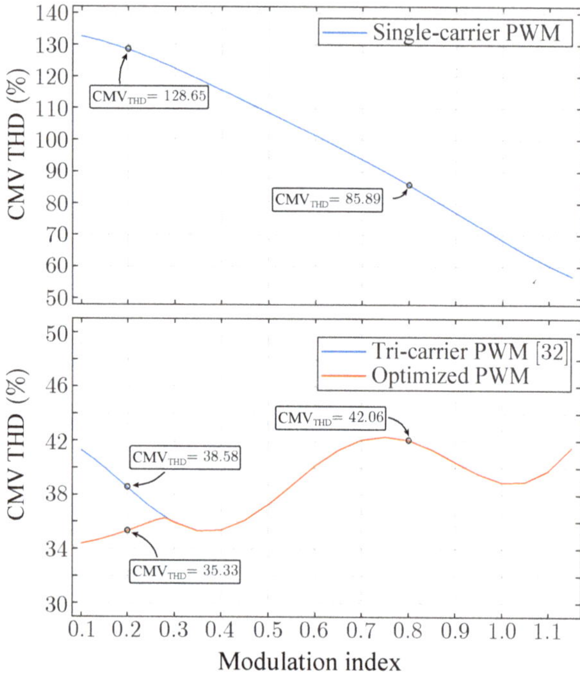

Figure 7. Evaluation of the CMV_{THD} with a modulation index range from 0.1 to 1.15. (**top**) Conv tional sigle-carrier PWM (**bottom**) Tricarrier PWM [33] and the Optimized PWM.

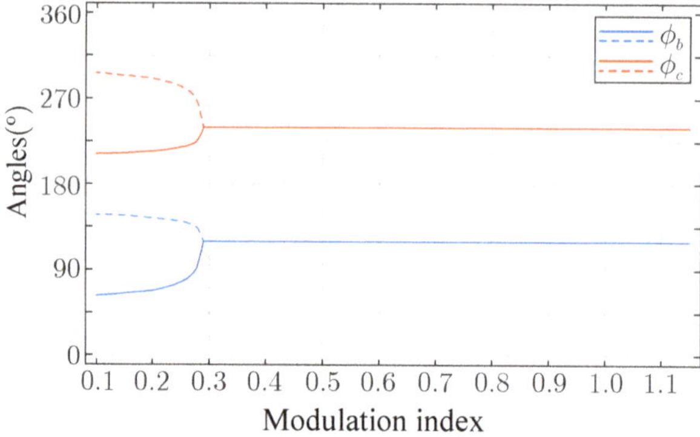

Figure 8. Carrier phase-displacement angles to achieve the minimization of the CMV_{THD} ir modulation index range from 0.1 to 1.15.

6. Experimental Results

In order to check the effectiveness of the PWM techniques and to validate the CM harmonic analysis performed in previous sections, the down-scaled laboratory expe mental setup shown in Figure 9 has been used. The experimental prototype consists o traditional three-phase two-level inverter connected to a 750 W three-phase PMSM wi main parameters summarized in Table 1. On the other hand, the PMSM parasitic comp nents are addressed in Figure 10. The controller and modulation schemes are implement using the rapid prototyping real-time platform PLECS RT box [36].

Table 1. Multi-phase PMSM parameters.

Parameters	Values
Winding resistance [Ω]	0.901
Winding inductance [mH]	6.552
Back-EMF coefficient K_E [V/rpm]	0.0227
Moment of inertia J [g/m^2]	0.12
Pole pair number	4
Voltage [V]	220
Nominal torque [Nm]	2.4
Nominal speed [rpm]	3000

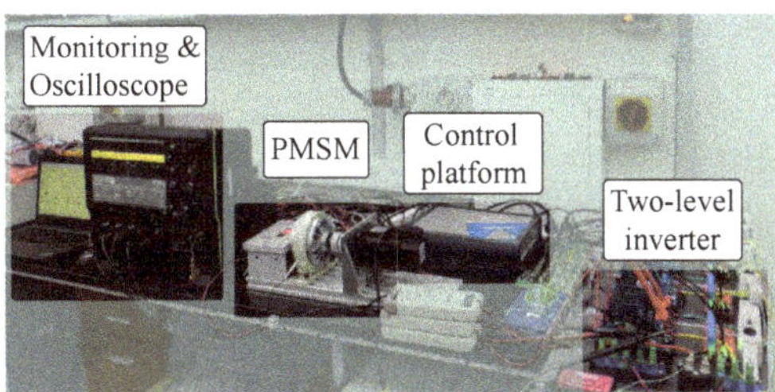

Figure 9. Down-scaled 750 W three-phase PMSM laboratory prototype.

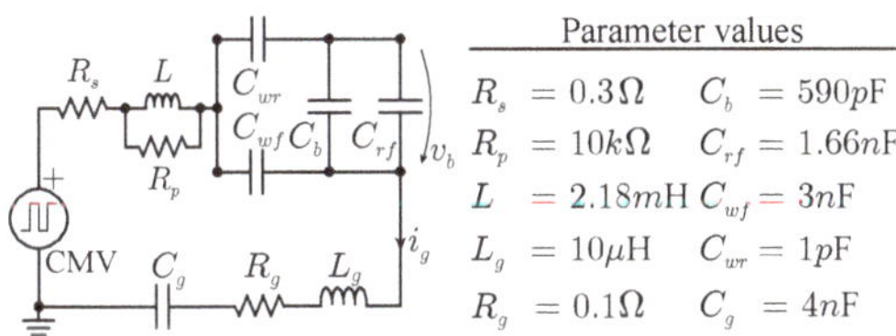

Figure 10. Parasitic components of the three-phase PMSM.

Figure 11 shows the PMSM control and modulation strategies implemented in the experimental setup. In the control strategy, the PMSM is driven by a speed and current double closed control loop [37]. In the modulation stage, the three PWM methods ($[\phi_a, \phi_b, \phi_c]$ equal to $[0,0,0]^\circ$, $[0,120,240]^\circ$ and $[0, \phi_{b_{opt}}, \phi_{c_{opt}}]^\circ$) are tested to check the influence on the VSD CMV, the phase currents as well as the induced shaft voltage.

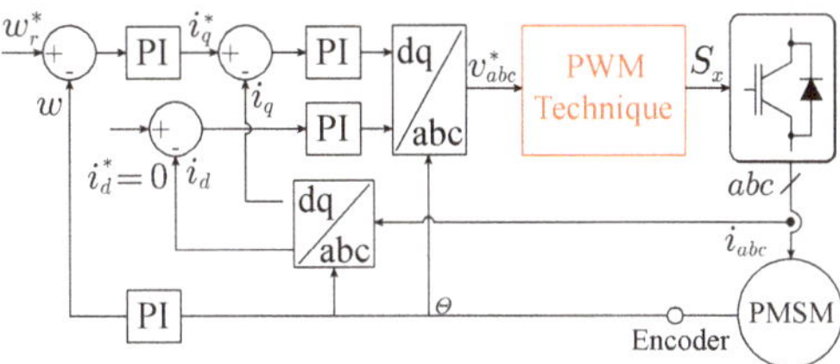

Figure 11. Controller scheme implemented in the experimental setup.

On one hand, the operation of the PMSM is analyzed in terms of the CMV generated by VSD applying the PWM method with the three set of carrier phase-displacement angles with $f_c = 4$ kHz and rotational speed equal to 25 rpm, that corresponds to apply a modulation index equal to 0.2. As it can be observed, applying the carrier phase-displacement PWM technique with $[\phi_a, \phi_b, \phi_c]$ equal to $[0,120,240]^\circ$, the peak-to-peak CMV and the CMV_{THD} are reduced compared with the single-carrier PWM method (see Figure 12c and Figure 12d, respectively). Applying this solution, the CMV harmonic content located at the switching frequency, that is $m = 1$ and $n = 0$ has been eliminated as expected. Howev the non-negligible harmonic component shown in Figure 12d surrounding f_c correspon with coefficients $m = 1$ and $n = 2$, what is expected from the analysis described in Figure In order to minimize the CMV_{THD}, the optimal values of the carrier phase-displaceme angles are applied and the corresponding results are summarized in Figure 12e,f. It ca be observed that this solution also achieves a reduction of the peak-to-peak CMV but th CMV_{THD} is reduced as well compared with the previous PWM methods as it was expecte observing Figure 6b.

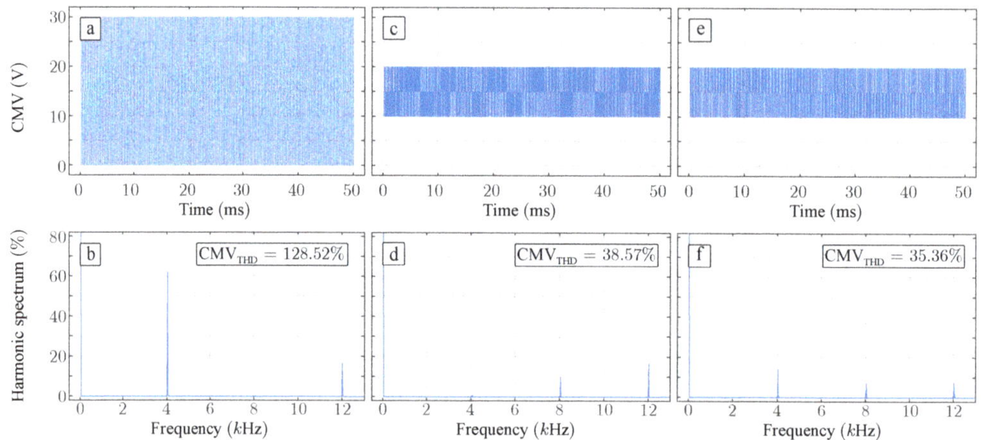

Figure 12. Results applying the traditional PWM method with $[\phi_a, \phi_b, \phi_c] = [0, 0, 0]^\circ$ (**a**) CMV generated in the PMSM (**b**) CMV harmonic spectrum. Results applying the PWM method with $[\phi_a, \phi_b, \phi_c] = [0, 120, 240]^\circ$ (**c**) CMV generated in the PMSM (**d**) CMV harmonic spectrum. Results applying the PWM method with $[\phi_a, \phi_b, \phi_c] = [0, \phi_{b_{opt}}, \phi_{c_{opt}}]^\circ$ (**e**) CMV generated in the PMSM (**f**) CMV harmonic spectrum.

In order to show another result, in Figure 13 is summarized the operation of the VS applying the PWM method with $[\phi_a, \phi_b, \phi_c]$ equal to $[0, 0, 0]^\circ$ and $[0, 120, 240]^\circ$ considerin $f_c = 4$ kHz, a speed reference equal to 250 rpm and a dc-link equal to 30 V. It is impo tant to notice that this operational point corresponds to a modulation index equal to 0.8 and, therefore, (as it was addressed in Figure 7 and Figure 8) the optimal carrier phas displacement angles are equal to the Kimball's solution [33], that is $[0, 120, 240]^\circ$. Fro Figure 13, it is clear that the application of the optimal values of carrier phase-displacemen angles leads to reduce both the peak-to-peak CMV and the CMV_{THD}. This experiment also used to evaluate the operation of the PMSM in terms of the phase currents qualit considering the PWM approaches. The results are reported in Figure 14. In Figure 14a th three-phase currents using the PWM method with $[\phi_a, \phi_b, \phi_c]$ equal to $[0, 0, 0]^\circ$ is show The harmonic spectrum of the phase a is represented in Figure 14b. In the same way, th three-phase currents as well as the harmonic spectrum of the phase a using the PWM technique with $[\phi_a, \phi_b, \phi_c]$ equal to $[0, 120, 240]^\circ$ are shown in Figure 14c and Figure 14 respectively. As it can be observed, the harmonic distortion of the phase currents appl ing the PWM technique with optimal carrier phase-displacement angles experiences a increasing. The normalized i_a current THD using the PWM method with $[\phi_a, \phi_b, \phi_c]$ equ to $[0, 0, 0]^\circ$ and $[0, 120, 240]^\circ$ are 3.86% and 4.86%, respectively. However, it can be seen th the magnitude of this degradation is not significant because it is the range of miliamper while the fundamental current component is 1.27 A.

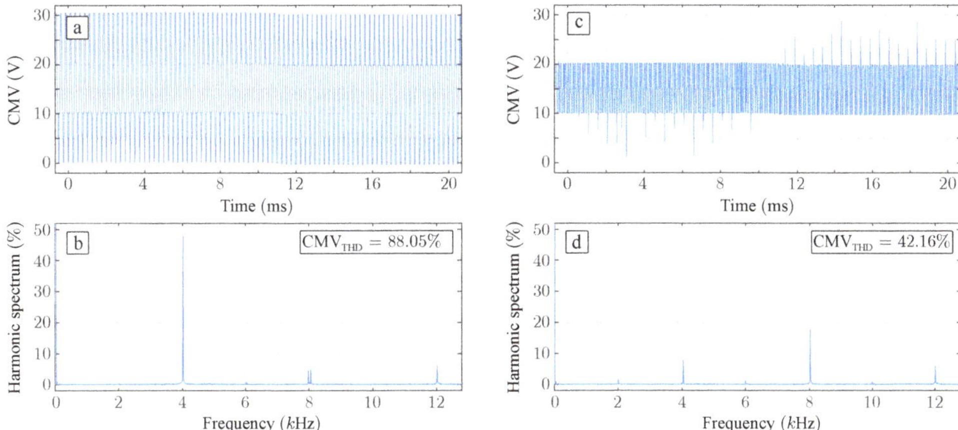

Figure 13. Results applying the traditional PWM method with $[\phi_a, \phi_b, \phi_c] = [0, 0, 0]^\circ$ (**a**) CMV generated in the PMSM (**b**) CMV harmonic spectrum. Results applying the PWM method with $[\phi_a, \phi_b, \phi_c] = [0, 120, 240]^\circ$ (**c**) CMV generated in the PMSM (**d**) CMV harmonic spectrum.

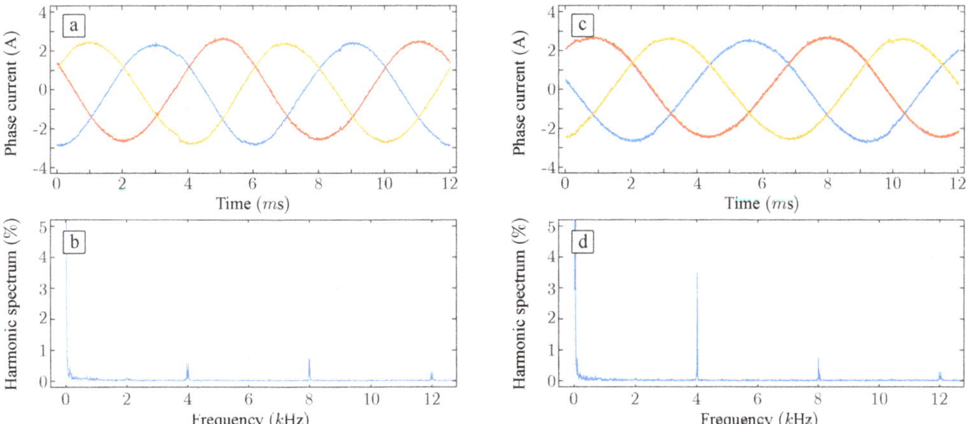

Figure 14. Results applying the traditional PWM method with $[\phi_a, \phi_b, \phi_c] = [0, 0, 0]^\circ$ (**a**) Phase currents (**b**) Harmonic spectrum of phase a current. Results applying the PWM method with $[\phi_a, \phi_b, \phi_c] = [0, 120, 240]^\circ$ (**c**) phase currents (**d**) Harmonic spectrum of phase a current.

These experiments have been carried out considering different operational conditions testing the performance of the methods with several angular speed values of the PMSM. The CMV_{THD} as well as the THD of the phase a current generated by the VSD are summarized in Table 2. It can be concluded that, using the PWM method with $[\phi_a, \phi_b, \phi_c]$ equal to $[0, 120, 240]^\circ$, the CMV harmonic mitigation is significant. On the other hand, a slight degradation of the phase currents occurs but this drawback can be considered acceptable if the design parameters are defined by other constraints. Indeed, the inductance of the machine depends on the mechanical requirements and on the voltage/current rating, whereas the switching frequency usually has a lower limit due to the control bandwidth. For many applications in industrial drives, which present high value of the machine inductance, the increase in the current THD caused by applying the carrier phase-displacement angle PWM method would be negligible.

Table 2. CMV_{THD} and normalized THD of the phase a current considering up to 13 kHz using the PWM with $[\phi_a, \phi_b, \phi_c]$ equal to $[0,0,0]^\circ$, $[0,120,240]^\circ$ and $[0, \phi_{b_{opt}}, \phi_{c_{opt}}]^\circ$.

		CMV_{THD} [%]			i_a THD [%]		
Speed [rpm]	**M [p.u.]**	**[0,0,0]°**	**[0,120,240]°**	**$[0, \phi_{b_{opt}}, \phi_{c_{opt}}]^\circ$**	**[0,0,0]°**	**[0,120,240]°**	**$[0, \phi_{b_{opt}}, \phi_{c_o}$**
25	0.20	128.65	38.58	35.33	1.90	6.25	5.29
50	0.40	121.38	35.77		2.74	5.76	
100	0.54	113.32	36.38		4.17	5.27	
150	0.66	105.50	39.14		3.21	5.16	
200	0.73	97.37	41.58		3.62	4.46	
250	0.84	88.05	42.16		3.71	5.02	

Finally, as it is mentioned in previous sections, the presence of high CMV in the V leads to the degradation of the motor bearing because of the corresponding high sh voltages and leakage currents. The shaft voltage present in the system applying the PW technique with $[\phi_a, \phi_b, \phi_c]$ equal to $[0,0,0]^\circ$ as well as those obtained applying $[0,120,24$ and $[0, \phi_{b_{opt}}, \phi_{c_{opt}}]^\circ$ have been measured. The experimental results with rotational spe equal to 25 rpm and 250 rpm are represented in Figure 15 and Figure 16, respectively. it can be observed from both experiments and their spectra, the shaft voltage has be reduced using the optimal carrier phase-displacement angles. This fact confirms th the use of this modulation method mitigates the motor bearings damage extending t machine lifetime.

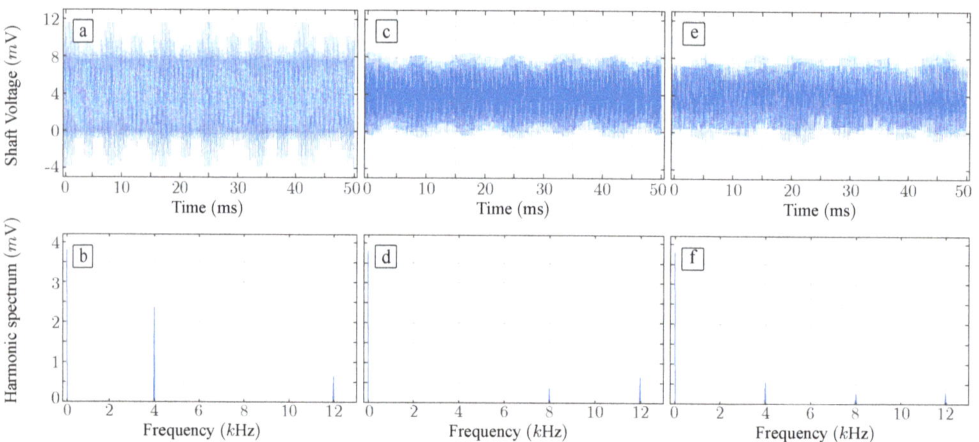

Figure 15. PMSM shaft voltage with rotational speed equal to 25 rpm (corresponding modulation index 0.2) using the PWM method with carrier phase-displacement angles $[\phi_a, \phi_b, \phi_c]$ equal to (**a**) $[0,0,0]^\circ$ (**c**) $[0,120,240]^\circ$ (**e**) $[0, \phi_{b_{opt}}, \phi_{c_{opt}}]^\circ$. (**b**,**d**,**f**) are their respective harmonic spectra.

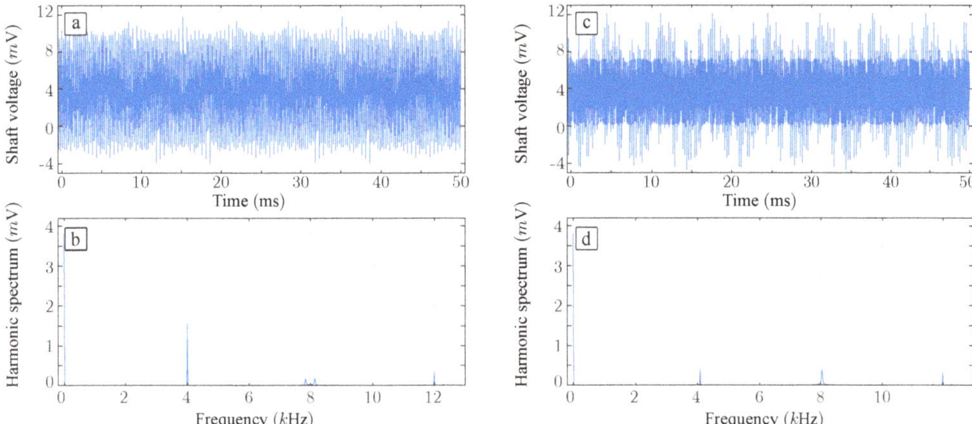

Figure 16. PMSM shaft voltage with rotational speed equal to 250 rpm (corresponding modulation index 0.84) using the PWM method with carrier phase-displacement angles $[\phi_a, \phi_b, \phi_c]$ equal to (**a**) $[0, 0, 0]^\circ$ (**c**) $[0, \phi_{b_{opt}}, \phi_{c_{opt}}]^\circ = [0, 120, 240]^\circ$. (**b**,**d**) are their respective harmonic spectra.

7. Conclusions

Motor drives are essential players in the current power electronics and power system arena. The introduction of the VSDs was revolutionary because it meant a significant increase in performance and controllability of the power systems. However, as a major concern related to the VSDs operation, it has been demonstrated that the CMV deteriorates the motor bearings leading to a reduction of the machine lifetime.

This paper presents an analysis of a generalized PWM method with triangular carriers with phase-displacement angles in order to reduce the CMV harmonic content without the introduction of passive filtering elements and/or active devices for the canceler stage. An accurate mathematical model based on double Fourier series is provided to describe the CMV content. This mathematical description is used to determine the best carrier phase-displacement angles to reduce the CMV content. In the paper, it is demonstrated that, considering a three-phase voltage source inverter to operate as a motor drive, the use of three carriers with phase displacements equal to $[0, 120, 240]^\circ$ is not always the best option in order to both reduce the low-frequency CMV harmonic components and also achieve the reduction of the CMV total harmonic distortion. For low values of the modulation index (up to 0.3) the optimal carrier phase-displacement angles are obtained, while applying $[0, 120, 240]^\circ$ is the most proper solution for larger values of the modulation index.

The accuracy of the analysis and the effectiveness of the generalized carrier phase-displacement angle PWM method have been validated via experimental results in a down-scaled laboratory prototype.

Author Contributions: Conceptualization, A.M.A., J.I.L., V.G.M. and L.G.F.; methodology, A.M.A., V.G.M.; software, A.M.A.; validation, V.G.M., X.W. and S.V.; formal analysis, A.M.A., G.B., V.G.M.; resources, M.L. and L.G.F.; writing, original draft preparation, and writing, review and editing, all authors. All authors have read and agreed to the published version of the manuscript.

Funding: The authors gratefully acknowledge financial support provided by the Andalusian Government and the European Union under Projects P18-RT-1340 and SPARTAN-821381, respectively.

Conflicts of Interest: The authors declare no conflict of interest.

References

1. Bose, B.K. Global Energy Scenario and Impact of Power Electronics in 21st Century. *IEEE Trans. Ind. Electron.* **2013**, *60*, 2638–26
[CrossRef]
2. Buticchi, G.; Bozhko, S.; Liserre, M.; Wheeler, P.; Al-Haddad, K. On-Board Microgrids for the More Electric Aircraft-Technolo Review. *IEEE Trans. Ind. Electron.* **2019**, *66*, 5588–5599. [CrossRef]
3. Holmes, D.G.; Lipo, T.A. *Pulse Width Modulation for Power Converters: Principles and Practice*; Wiley-IEEE Press: Hoboken, NJ, US 2003. [CrossRef]
4. Yazdani, A.; Iravani, R. *Voltage-Sourced Converters in Power Systems: Modeling, Control, and Applications*; Wiley-IEEE Press: Hobok NJ, USA, 2010.
5. Leon, J.I.; Kouro, S.; Franquelo, L.G.; Rodriguez, J.; Wu, B. The Essential Role and the Continuous Evolution of Modulati Techniques for Voltage-Source Inverters in the Past, Present, and Future Power Electronics. *IEEE Trans. Ind. Electron.* **20** *63*, 2688–2701. [CrossRef]
6. Leon, J.I.; Vazquez, S.; Franquelo, L.G. Multilevel Converters: Control and Modulation Techniques for Their Operation ar Industrial Applications. *Proc. IEEE* **2017**, *105*, 2066–2081. [CrossRef]
7. Wilamowski, B.M.; Irwin, J.D. *Power Electronics and Motor Drives*; CRC Press: Boca Raton, FL, USA, 2017.
8. Zhou, K.; Wang, D. Relationship between space-vector modulation and three-phase carrier-based PWM: A comprehensi analysis [three-phase inverters]. *IEEE Trans. Ind. Electron.* **2002**, *49*, 186–196. [CrossRef]
9. Eaton, D.; Rama, J.; Hammond, P. Neutral shift [five years of continuous operation with adjustable frequency drives]. *IEEE I Appl. Mag.* **2003**, *9*, 40–49. [CrossRef]
10. Kalaiselvi, J.; Srinivas, S. Bearing Currents and Shaft Voltage Reduction in Dual-Inverter-Fed Open-End Winding Inducti Motor With Reduced CMV PWM Methods. *IEEE Trans. Ind. Electron.* **2015**, *62*, 144–152. [CrossRef]
11. Mutzw, A. Thousands of hits: On inverter-induced bearing currents, related work, and the literature. *Elektrotechnik u Informationstechnik* **2011**, *128*, 382–388. [CrossRef]
12. Skibinski, G.L.; Kerkman, R.J.; Schlegel, D. EMI emissions of modern PWM AC drives. *IEEE Ind. Appl. Mag.* **1999**, *5*, 47–
[CrossRef]
13. Isomura, Y.; Yamamoto, K.; Morimoto, S.; Maetani, T.; Watanabe, A.; Nakano, K. Study of the Further Reduction of Shaft Volta of Brushless DC Motor With Insulated Rotor Driven by PWM Inverter. *IEEE Trans. Ind. Appl.* **2014**, *50*, 3738–3743. [CrossRef]
14. Jiang, D.; Chen, J.; Shen, Z. Common mode EMI reduction through PWM methods for three-phase motor controller. *CES Tra Electr. Mach. Syst.* **2019**, *3*, 133–142. [CrossRef]
15. Lee, S.; Park, J.; Jeong, C.; Rhyu, S.; Hur, J. Shaft-to-Frame Voltage Mitigation Method by Changing Winding-to-Rotor Parasit Capacitance of IPMSM. *IEEE Trans. Ind. Appl.* **2019**, *55*, 1430–1436. [CrossRef]
16. Han, Y.; Lu, H.; Li, Y.; Chai, J. Analysis and Suppression of Shaft Voltage in SiC-Based Inverter for Electric Vehicle Applicatior *IEEE Trans. Power Electron.* **2019**, *34*, 6276–6285. [CrossRef]
17. Hava, A.M.; Un, E. A High-Performance PWM Algorithm for Common-Mode Voltage Reduction in Three-Phase Voltage Sour Inverters. *IEEE Trans. Power Electron.* **2011**, *26*, 1998–2008. [CrossRef]
18. Muralidhara, B.; Ramachandran, A.; Srinivasan, R.; Reddy, M.C. Experimental measurement of shaft voltage and bearing curre in an inverter fed three phase induction motor drive. In Proceedings of the 2011 3rd International Conference on Electroni Computer Technology, Kanyakumari, India, 8–10 April 2011; Volume 2, pp. 37–41. [CrossRef]
19. Song-Manguelle, J.; Schröder, S.; Geyer, T.; Ekemb, G.; Nyobe-Yome, J. Prediction of Mechanical Shaft Failures Due to Pulsatir Torques of Variable-Frequency Drives. *IEEE Trans. Ind. Appl.* **2010**, *46*, 1979–1988. [CrossRef]
20. Magdun, O.; Binder, A. High-Frequency Induction Machine Modeling for Common Mode Current and Bearing Volta Calculation. *IEEE Trans. Ind. Appl.* **2014**, *50*, 1780–1790. [CrossRef]
21. Flieh, H.M.; Totoki, E.; Lorenz, R.D. Dynamic Shaft Torque Observer Structure Enabling Accurate Dynamometer Transient Lo Measurements. *IEEE Trans. Ind. Appl.* **2018**, *54*, 6121–6132. [CrossRef]
22. Naik, R.; Nondahl, T.A.; Melfi, M.J.; Schiferl, R.; Wang, J.-S. Circuit model for shaft voltage prediction in induction motors fed b PWM-based AC drives. *IEEE Trans. Ind. Appl.* **2003**, *39*, 1294–1299. [CrossRef]
23. Choochuan, C. A survey of output filter topologies to minimize the impact of PWM inverter waveforms on three-phase A induction motors. In Proceedings of the 2005 International Power Engineering Conference, Singapore, 29 November–2 Decemb 2005; Volume 2005, pp. 1–544. [CrossRef]
24. Mei, C.; Balda, J.C.; Waite, W.P. Cancellation of common-mode Voltages for induction motor drives using active method. *IEE Trans. Energy Convers.* **2006**, *21*, 380–386. [CrossRef]
25. Di Piazza, M.C.; Tine, G.; Vitale, G. An Improved Active Common-Mode Voltage Compensation Device for Induction Mot Drives. *IEEE Trans. Ind. Electron.* **2008**, *55*, 1823–1834. [CrossRef]
26. Fan, F.; See, K.Y.; Banda, J.K.; Liu, X.; Gupta, A.K. Investigation and mitigation of premature bearing degradation in motor driv system. *IEEE Electromagn. Compat. Mag.* **2019**, *8*, 75–81. [CrossRef]
27. Jayaraman, K.; Kumar, M. Design of Passive Common-Mode Attenuation Methods for Inverter-Fed Induction Motor Drive Wi Reduced Common-Mode Voltage PWM Technique. *IEEE Trans. Power Electron.* **2020**, *35*, 2861–2870. [CrossRef]
28. Tallam, R.M.; Kerkman, R.J.; Leggate, D.; Lukaszewski, R.A. Common-Mode Voltage Reduction PWM Algorithm for AC Drive *IEEE Trans. Ind. Appl.* **2010**, *46*, 1959–1969. [CrossRef]

Cacciato, M.; Consoli, A.; Scarcella, G.; Testa, A. Reduction of common-mode currents in PWM inverter motor drives. *IEEE Trans. Ind. Appl.* **1999**, *35*, 469–476. [CrossRef]
Adabi, J.; Boora, A.A.; Zare, F.; Nami, A.; Ghosh, A.; Blaabjerg, F. Common-mode voltage reduction in a motor drive system with a power factor correction. *IET Power Electron.* **2012**, *5*, 366–375. [CrossRef]
Espina, J.; Ortega, C.; de Lillo, L.; Empringham, L.; Balcells, J.; Arias, A. Reduction of Output Common Mode Voltage Using a Novel SVM Implementation in Matrix Converters for Improved Motor Lifetime. *IEEE Trans. Ind. Electron.* **2014**, *61*, 5903–5911. [CrossRef]
Espina, J.; Ortega, C.; de Lillo, L.; Empringham, L.; Balcells, J.; Arias, A. Common-Mode Voltage Mitigation Technique in Motor Drive Applications by Applying a Sampling-Time Adaptive Multi-Carrier PWM Method. *IEEE Access* **2021**, *9*, 56115-56126. [CrossRef]
Kimball, J.W.; Zawodniok, M. Reducing Common-Mode Voltage in Three-Phase Sine-Triangle PWM With Interleaved Carriers. *IEEE Trans. Power Electron.* **2011**, *26*, 2229–2236. [CrossRef]
Alcaide, A.M.; Wang, X.; Yan, H.; Leon, J.I.; Monopoli, V.G.; Buticchi, G.; Vazquez, S.; Liserre, M.; Franquelo, L.G. Common-Mode Voltage Mitigation of Dual Three-Phase Voltage Source Inverters in a Motor Drive Application. *IEEE Access* **2021**, *9*, 67477–67487. [CrossRef]
Shen, Z.; Jiang, D.; Liu, Z.; Ye, D.; Li, J. Common-Mode Voltage Elimination for Dual Two-Level Inverter-Fed Asymmetrical Six-Phase PMSM. *IEEE Trans. Power Electron.* **2020**, *35*, 3828–3840. [CrossRef]
Plexim RT Box Rapid Prototype Platform. Available online: https://www.plexim.com/products/rt_box (accessed on 6 April 2021).
Zheng, B.; Xu, Y.; Wang, G.; Yu, G.; Yan, H.; Wang, M.; Zou, J. Carrier frequency harmonic suppression in dual three-phase permanent magnet synchronous motor system. *IET Electr. Power Appl.* **2019**, *13*, 1763–1772. [CrossRef]

Article

Analysis and Implementation of Multilevel Inverter for Full Electric Aircraft Drives

Vidhi Patel *,†, Concettina Buccella † and Carlo Cecati †

Department of Information Engineering, Computer Science and Mathematics, University of L'Aquila, and DigiPower srl, 67100 L'Aquila, Italy; concettina.buccella@univaq.it (C.B.); carlo.cecati@univaq.it (C.C.)
* Correspondence: vidhimanilal.patel@graduate.univaq.it
† These authors contributed equally to this work.

Received: 16 October 2020; Accepted: 18 November 2020; Published: 22 November 2020

Abstract: In modern aircrafts, hydraulic or pneumatic actuators have been already replaced with electric counterparts, but the advancement of the inverter and motor technology has made possible that the propulsion system can be powered by electrical sources. These high power requirements can not be efficiently fulfilled by using a typical two level converter; the multi-level converter represents a suitable solution for this application. This paper presents a cascaded H-bridge, a 9-level permanent magnet synchronous motor drive for full electric aircrafts. Harmonic analysis is presented considering different levels of a multi-level inverter. Simulation results and experimental validation with a test-rig confirms the accuracy of the proposed system.

Keywords: full electric aircraft (FEA); cascaded H-bridge (CHB); multi-level inverter; permanent magnet synchronous motor (PMSM); total harmonic distortion (THD)

1. Introduction

Due to increasing power efficiency demand, reduced high operating cost and most importantly reduced air pollution, mobility is changing drastically and great attention is focused on the electrification in transportation. To accomplish this, power electronics (PE) and electric machines are needed and required to be much more reliable, efficient, compact, quieter and with high power density. The electrically propelled vehicle is dependent majorly on the PE to meet challenges like specific power, energy storage capabilities, thermal management of the system and most importantly safety [1]. In the conventional aircraft, actuation, de-icing, air conditioning, cabin pressurization, fuel pumping and other systems, are powered by hydraulic, pneumatic and mechanical actuators and propulsion system is powered by kerosene. In a full electric aircraft (FEA), all power systems, including propulsion are powered by electrical sources. This leads to increased dominance of power electronics, which opens the research window for the cleaner sky [2].

Currently, there are more than two hundred electric aircraft demonstrators approaching different methodologies which grew the number of electrically driven aircraft development approximately by 30% in last year [3]. Some of them include CityAirbus 4-seater all-electric eVTOL, which is comprised of eight 100 kW electric motors with a DC bus of 800 V [4]. Airbus is also exploring its wings on large aircrafts like series hybrid e-fan X that features 2 MW power converter having a DC bus of 3 kV [5]. In this race, a hybrid electric propulsion system by Bell also developed a system with six 100 kW motors to produce a power of 600 kW [6]. The required on-board electric power of a Boeing 787 is provided by four 250 kVA generators and two 225 kVA auxiliary power units (APU) [2,7]. In the existing Boeing 737, the electrical system capacity is 100 kW, which in the recent model is increased over 1 MW. Generally speaking, if an electric powertrain is adopted, very high power and medium voltage drives become mandatory and cascaded H-bridge (CHB) multi-level inverters (MLIs) represent an attractive solution for the numerous advantages in terms of high efficiency, power quality and because of the use of low rated power devices, including the novel wide band gap (WBG) devices like gallium nitride (GaN) and silicon carbide (SiC). While conventional two level inverters suffer from high dv/dt, high blocking voltage across power device, common mode voltage, etc., which may lead to issues such as overheating and failures of motor winding, MLIs offer sinusoidal like output voltage profile, reduced dv/dt and consequently lower stress on power devices and on the motor, redundancy and reduced total harmonic distortion (THD). It can also handle high power with low switching frequency and thus lower switching losses, low common mode voltage, low stress in the motor bearings and in turn increasing the life of the motor [8–10].

The modern wide band gap (WBG) devices will play a crucial role in the race of electrification of high efficient PE converters for full electric aircrafts, bearing higher temperature. There are various works in the literature which demonstrate the merits of WBG for FEA. The paper [11], for instance, demonstrates a 2-level inverter of 50 kW using SiC devices to achieve high power density of the aircraft, resulting in saving of energy. On the other hand, the modern WBG operate at frequency higher than typical Si devices and this could increase electromagnetic interference (EMI) giving rise to higher di/dt and dv/dt ratios [1]. Hence, an alternative approach both at system and component levels has to be adopted. Nonetheless, SiC and GaN devices are still a quite immature technology with high cost. Their use requires re-design of the whole power converter system including gate drivers and so on.

A megawatt scale medium voltage three level active neutral point clamped power converter is proposed for high efficiency and high density aircraft hybrid propulsion systems in [12]. The paper [13] gives the detailed analysis for different modulation techniques and the LCL filter requirements using a five-level CHB inverter. The space agencies like NASA are exploiting the possibility of MLI for FEA application. For the electrically propelled aircraft research program, General Electric is considering a 1 MW, 3-level inverter with 2.4 kV DC-link voltage [14]. In the powertrain applications, electric machines represent the heart. In 2017, Airbus replaced one of the four jet engines by a 2 MW electric motor [5]. Numerous kinds of machines are being used for the power generation of the engine, such as switch reluctance [15], permanent magnet (PM) [16] and special designed induction machines [17]. Very often, machines are multi-phase with 5, 6 or even more phases. The literature shows that PM machine has attractive features when compared to others to fit the requirement of FEA, because it offers high power density [15], efficiency, reliability and ruggedness. The performance comparison in [18], among many others, proves their superiority.

This paper proposes a multi-level CHB inverter to feed three phase permanent magnet synchronous motor (PMSM) for FEA powertrain applications. The study of multi-phase and fault tolerant solutions, more suitable in terms of reliability, is in progress and will be the subject of future papers. Section 2 presents the mathematical formulation of the whole system. In Section 3, advantages of MLI for FEA drives are discussed. The study is presented for l-level CHB MLI topologies, that justifies that the 9-level one gives the best trade-off between performance and cost [19]. Therefore, the 9-level inverter is considered for the explanation of MLI drive, but analysis is presented for 2, 5, 7 and 9-level inverter configurations. Simulated results and experimental implementation are presented in Section 4. Section 5 gives some conclusions.

2. CHB 9-Level Inverter Powertrain

2.1. Operation of 9-Level Inverter

CHB 9-level topology consists of four H-bridge modules per phase supplied with equal dc voltage sources, i.e., $V_{dc}/4$. Considering the phase A, the semiconductor devices are denoted as S_{1a}, S_{2a}, ... up to S_{16a}, as shown in Figure 1. In each H-bridge, switches of the same leg such as S_{1a}, S_{4a} and S_{5a}, S_{8a} and so on, are operated in complimentary fashion. In a similar manner, phase B and C are configured, 120° and 240° delayed from phase A, respectively. To obtain the staircase output waveform, the switching states are applied as shown in Table 1. To reduce losses, the switching states are selected in a manner that only one switch has to be turned on, when the level is changed. The adopted modulation technique is the level-shifted carrier (LSC)-based PWM technique, with triangular wave C_j, $j = 1, 2, ..., 8$ as carrier with frequency equal to 10 kHz, as in Figure 2. This modulation technique eliminates the sector identification and complex timing calculations, which are necessary for the implementation of space vector PWM (SVM) technique. To obtain the same reference voltage waveforms of SVM, a third harmonic component has been introduced in order to increase the linear modulation range [20].

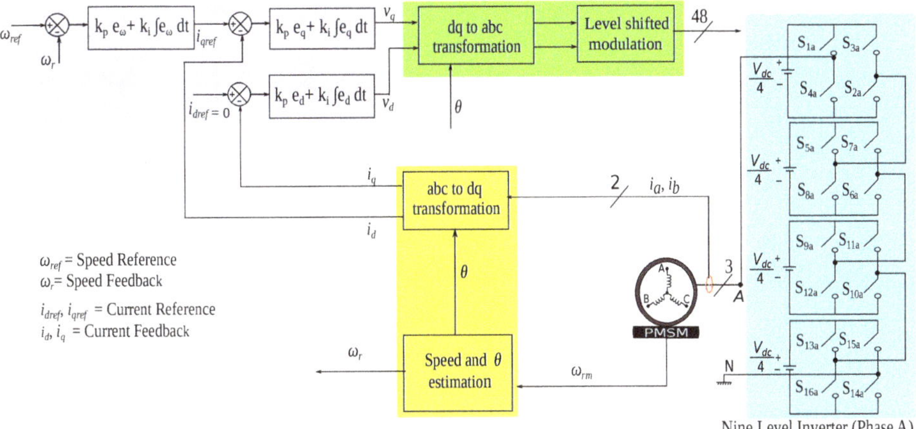

Figure 1. Block diagram of the 9-level inverter drive.

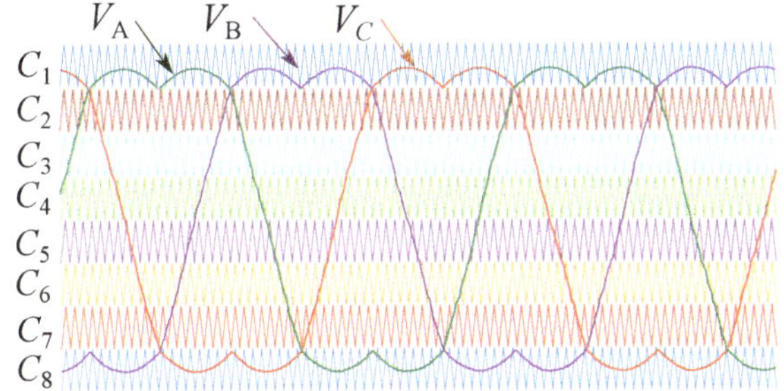

Figure 2. Carrier and modulating waveforms for LSCPWM.

Table 1. Nine voltage levels and their switching states.

V_{xN} *	Level	Switching States							
		S_{1a}	S_{3a}	S_{5a}	S_{7a}	S_{9a}	S_{11a}	S_{13a}	S_{15a}
V_{dc}	8	1	0	1	0	1	0	1	0
$3V_{dc}/4$	7	1	0	1	0	1	0	0	0
$2V_{dc}/4$	6	1	0	1	0	0	0	0	0
$V_{dc}/4$	5	1	0	0	0	0	0	0	0
0	4	0	0	0	0	0	0	0	0
$-V_{dc}/4$	3	0	1	0	0	0	0	0	0
$-2V_{dc}/4$	2	0	1	0	1	0	0	0	0
$-3V_{dc}/4$	1	0	1	0	1	0	1	0	0
$-V_{dc}$	0	0	1	0	1	0	1	0	1

* $x = A, B, C$.

2.2. Operational Principle of PMSM

In this study, aiming to investigate the behavior of multi-level PMSM drive, the conventional field oriented control (FOC) technique is implemented. Mathematical model of PMSM is formulated in *d-q* co-ordinates as

$$\begin{bmatrix} v_d \\ v_q \end{bmatrix} = \begin{bmatrix} R_s + L_d\rho & -\omega_r L_q \\ \omega_r L_d & R_s + L_q\rho \end{bmatrix} \begin{bmatrix} i_d \\ i_q \end{bmatrix} + \begin{bmatrix} 0 \\ \omega_r\phi \end{bmatrix}. \tag{1}$$

where v_d, v_q and i_d, i_q are stator voltages and currents in *d-q* axis, respectively, ρ is the differential operator, R_s is stator resistance, ω_r and ϕ are electrical angular speed and flux linkage of PMSM, respectively. L_d and L_q are stator inductances which have the same value L, a surface mounted type of PMSM is used. This type of motor has a small rotor diameter with low inertia and hence, finer dynamic performance. The electromagnetic torque T_e is given by

$$T_e = \frac{3}{2}\mathrm{P}\,\phi i_q. \tag{2}$$

where P is number of pole pairs of the motor. The mechanical speed ω_{rm} is

$$\omega_{rm} = \frac{1}{J}\int (T_e - T_l - B\omega_{rm})dt. \tag{3}$$

where T_l, B, J are load torque, viscous friction coefficient and moment of inertia of the motor, respectively. The electrical speed of the machine can be obtained by using $\omega_r = \mathrm{P}\,\omega_{rm}$. The electrical angle of the rotor θ is estimated from incremental encoder. Figure 1 shows the two cascaded control loops, outer is speed loop and the inner one is the current loop. For the speed control of drive, the speed error is generated by comparing the speed reference ω_{ref} with the obtained speed ω_r. This error is processed by speed PI controller, returning the current reference i_{qref}. The i_q is subtracted from i_{qref}, generating an error that is handled by the q PI controller, giving v_q stator voltage. The FOC method decouples the d and q axis, referring d to flux control and q to torque control. The stator voltage v_d is estimated from the error generated by comparing flux controller reference, i.e., zero and i_d. The v_d and v_q are converted to reference voltages V_A, V_B and V_C, for the comparison with carrier waves as explained in Section 2.1. The transformations from *dq* to *abc* and vice versa are performed using Clarke's and Park's transformations.

3. Analysis of CHB Multi-Level for FEA

In this section, the advantages obtained by using a CHB multi-level inverter in terms of THD% and torque ripple are discussed in detail. The harmonic mitigation is a great concern for power quality of converters as well as of overall FEA system. THD% is calculated using the following formula

$$\mathrm{THD} = \frac{\sqrt{\sum_{n=2,3,\ldots}^{49} H_n^2}}{H_1} \tag{4}$$

where H_n is the amplitude of the nth harmonic. For harmonic spectrum, in the y-axis, the amplitude of the n-harmonic H_n with respect to the fundamental H_1, in percentage value, is indicated. The considered fundamental frequency is 300 Hz and modulation index (MI) is 0.8. Figures 3a,b and 4a,b show the harmonic spectrum of output voltage of 2, 5, 7 and 9-level inverters, respectively.

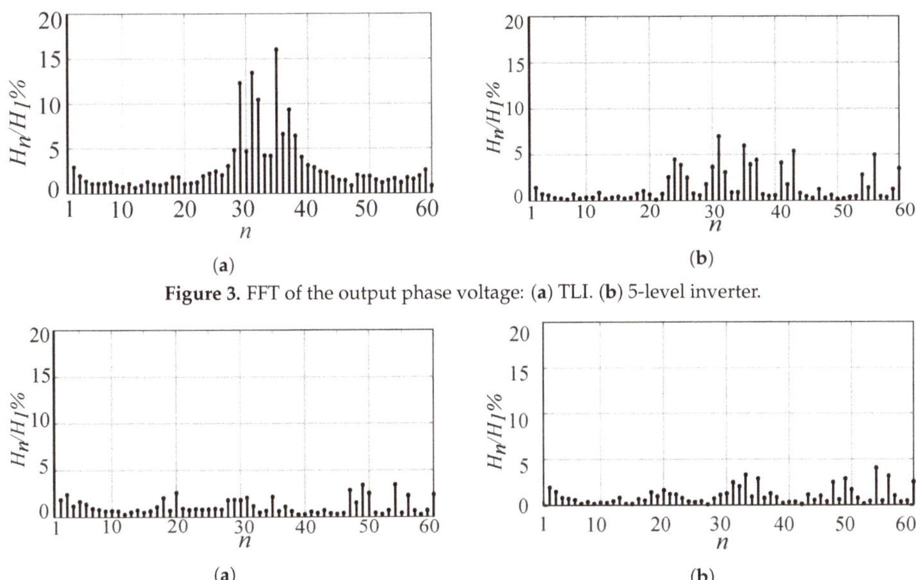

Figure 3. FFT of the output phase voltage: (a) TLI. (b) 5-level inverter.

Figure 4. FFT of the output phase voltage: (a) 7-level inverter. (b) 9-level inverter.

It can be observed that the 2-level inverter has higher amplitude of dominant harmonics 31st and the 35th than that of the 9-level inverter. The harmonic spectrum of line-line voltage for 2, 5, 7 and 9-level inverter configurations without using a filter, are presented as shown in Figures 5a,b and 6a,b, respectively. It can be seen that in the 9-level inverter, all harmonics do not have significant amplitude at the considered frequency range. The current harmonic spectrum for all configurations are also presented as shown in Figures 7a,b and 8a,b, respectively. This signifies that 9-level configuration requires a very light filter with respect to the 2-level inverter configuration. To obtain the same performances of 9-level configuration, a 2-level inverter must be used with higher switching frequency. In this case, the first not mitigated harmonic displaces to high order, but also switching, iron and copper losses increase.

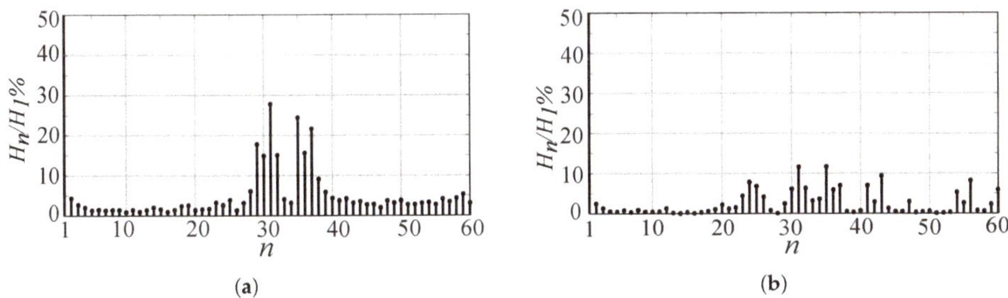

Figure 5. FFT of the output line-line voltage: (a) TLI. (b) 5-level inverter.

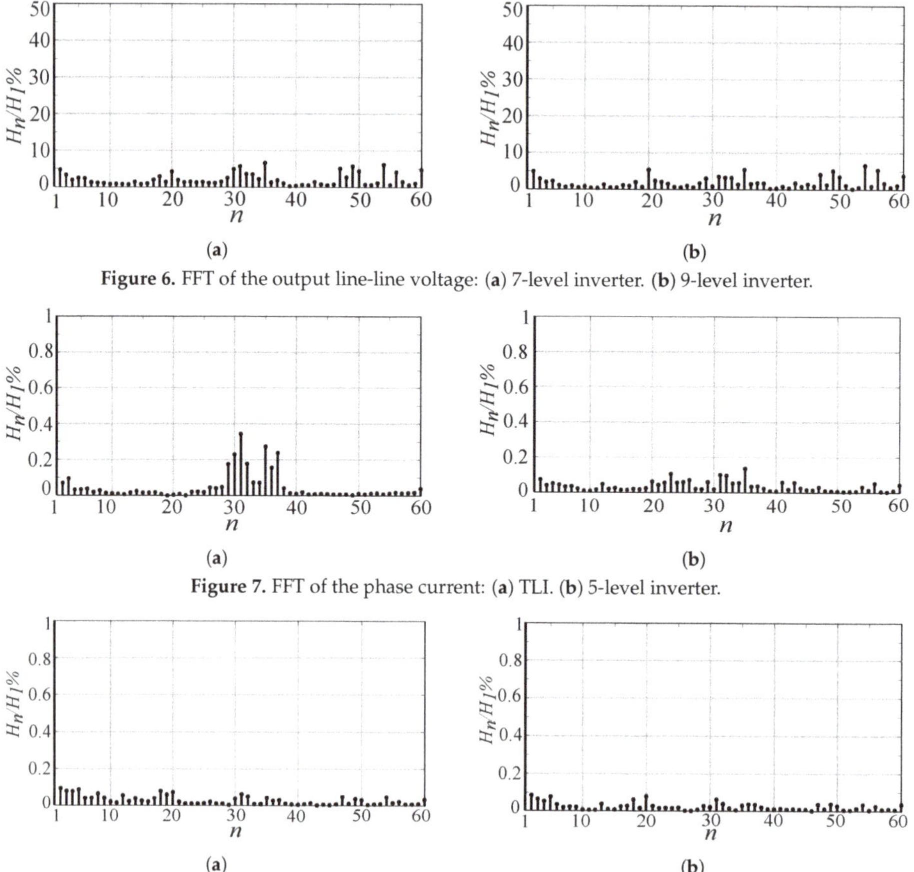

Figure 6. FFT of the output line-line voltage: (**a**) 7-level inverter. (**b**) 9-level inverter.

Figure 7. FFT of the phase current: (**a**) TLI. (**b**) 5-level inverter.

Figure 8. FFT of the phase current: (**a**) 7-level inverter. (**b**) 9-level inverter.

An investigation about the use of the *l*-level inverter powertrain for FEA is performed for l = 2, 5, 7, 9, for different MIs of the drive. Figure 9a shows THD% as a function of MI. For MI = 1, THD% = 56% for the 2-level inverter and THD% = 8.29% for the 9-level inverter. For the 5 and 7-level inverters, THD% is 16.62% and 9.95%, respectively. For the 9-level inverter, at MI = 0.8 and 0.4, THD% = 8.05% and 19.05%, respectively. Considering the 9-level inverter, if MI decreases from 0.8 to 0.4, THD% increases from 8.05% to 19.05%.

The analysis of torque ripple as a function of MI, for 2, 5, 7 and 9-level inverters is shown in Figure 9b. The y-axis quantities represent the percentage torque values with respect to its nominal value. At MI = 0.8, torque ripple for 2, 5, 7 and 9-level inverters is 0.138, 0.109, 0.0871 and 0.06617, respectively and for MI = 0.2, it is 0.268, 0.1547, 0.099 and 0.073, respectively. These results show that the 9-level inverter gives better performance compared to other configurations in terms of THD and torque ripple. A conventional 2-level inverter contains 6 high rated switches, while 9-level configuration contains 48 switches, but since the DC-link voltage is shared among them, they can be low rated devices.

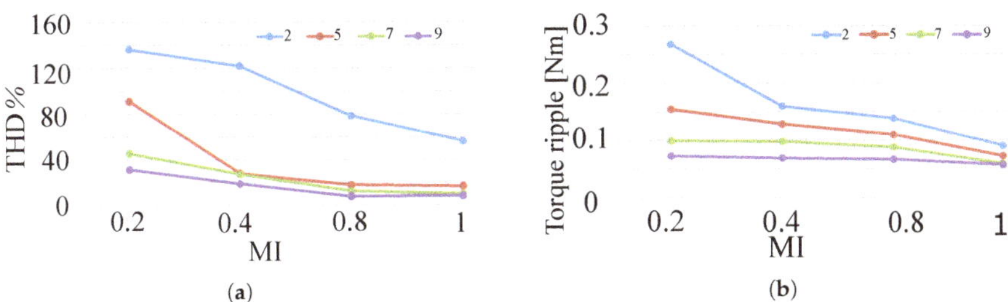

Figure 9. Comparison between 2, 5, 7 and 9-level inverter configurations: (**a**) total harmonic distortion (THD)%. (**b**) Torque ripple.

The efficiency of the drive is affected by high THD, torque ripple, switching and conduction losses, stress across the devices and this indirectly affects the life of the motor. The high torque ripple of the drive leads to vibrations, noise and wear of the machine. This not only damages the machine, but also the shaft and gearbox. The additional considered advantage of CHB MLI is that the total DC-link V_{dc} is used to create levels, so each power switch has to share 1/4 (for 9-level inverter) of the total DC voltage. The switches have to bear less voltage stress across them, reducing dv/dt. High dv/dt indirectly affects the longevity of motor by dielectric heating of the insulation material [21]. Therefore, the promising use of multi-level inverters for FEA could significantly increase the life of the motor [22]. This sharing of voltages also makes it easier to find the lower voltage devices with lower price, which may reduce overall production and maintenance cost. These results show that the 9-level inverter gives better performance compared to the other configurations in terms of THD and torque ripple. Conventional TLI contains 6 high rated switches, while the 9-level configuration contains 48 switches. Since the DC-link voltage is shared among them, they can be low rated devices.

4. Results

This section presents simulation and experimental results considering the 9-level inverter that feeds PMSM for FEA application. Considering the transition from 2-level to 5-level to 7-level to 9-level, the 8 carrier waveforms, the phase voltage and the speed are shown in Figure 10. Figure 11 shows simulated results of the phase voltage, the line-line voltage and the phase current at a steady state speed of 1200 RPM. In addition, the zoomed view of these waveforms are shown. Figure 12 shows simulated speed feedback that follows the reference when speed changes from 500 RPM to 3000 RPM. There is a small change in torque and phase currents before it reaches the steady state, as shown in Figure 12. In Figure 13, a variation of torque from 0.1 Nm to 1 Nm is considered; little variation of speed and currents is observed.

Figure 14 shows the block diagram of the hardware implementation of 9-level inverter powertrain. The CHB 9-level inverter powertrain includes 12 H-bridge modules communicating with the control board through dedicated SPI channels. Each H-bridge uses STW120NF10 MOSFETs (100 V, 120 A), and is rated for 3.5 kW; moreover, it includes a local Texas Instruments TMS320F28379D DSP, which is in charge of I/O and A/D conversion, but it is expected to assume the role of computational engine for fault detection, diagnosis and reconfiguration algorithms. It is shown in Figure 14. The same H-bridge is available with 1200 V, 35 A insulated gate bipolar transistors (IGBTs), thus allowing to obtain more than 200 kW and 2.4 kV output voltage, which is a suitable power level for light aircrafts. It is expected to replace the three phase motor with a six phase motor under development. The new configuration requires 24 H-bridges, hence the control board is capable to offer a 9-level redundant converter. The control board mainly consists of 5CEBA9F31 CYCLONE V FPGA, 10M16DAF484 MAX10 CPLD and TMS320F28379D DSP. The internal architecture of the control board is as shown in Figure 15. The whole system, i.e., H-bridges, control board and auxillary circuits, has been developed by DigiPower srl, an innovative SME based in L'Aquila, Italy.

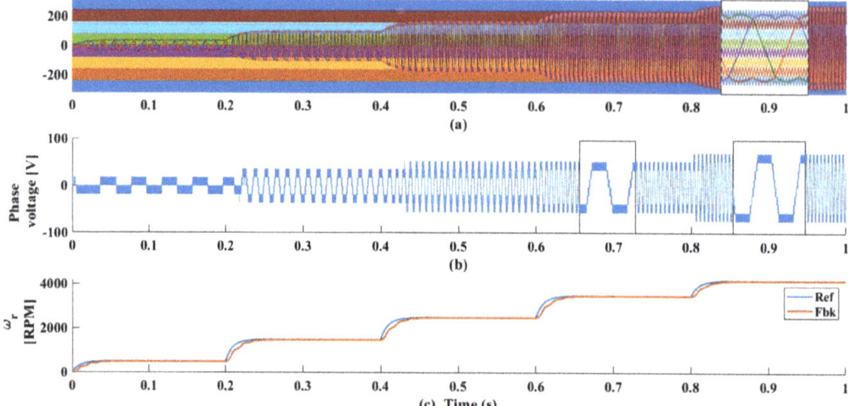

Figure 10. Simulation result obtained by using the 9-level inverter: (**a**) Eight carrier waveforms and modulating waveforms. (**b**) Phase voltage. (**c**) Speed reference (Ref) and speed feedback (Fbk).

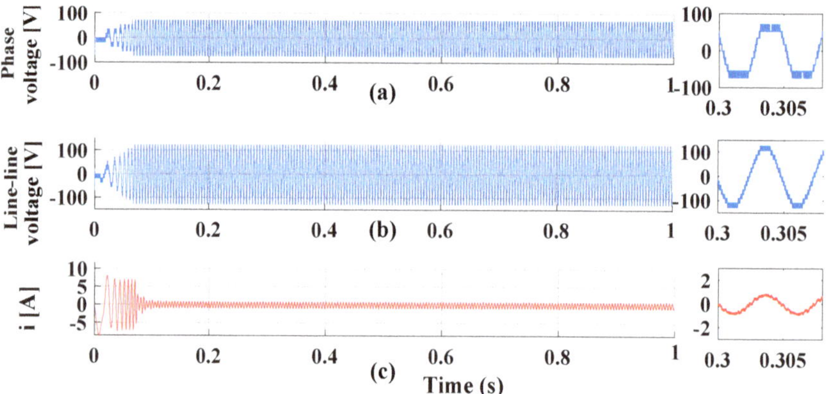

Figure 11. Simulation result of the 9-level inverter: (**a**) Phase voltage. (**b**) Line-line voltage. (**c**) Phase current.

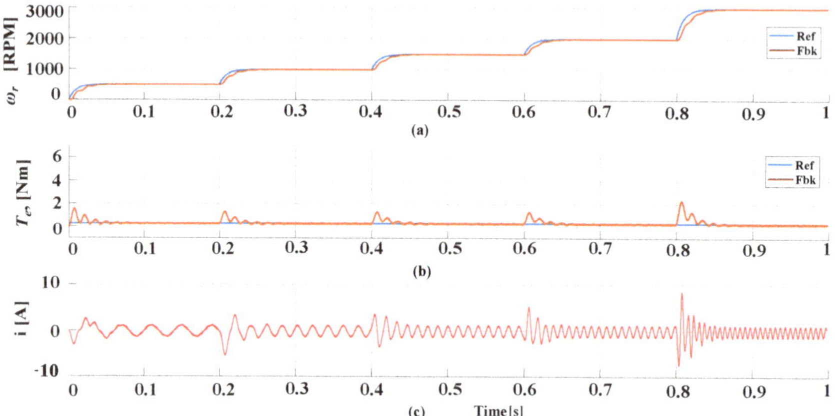

Figure 12. Simulation result of the 9-level inverter when the speed is varied: (**a**) Speed reference and its feedback. (**b**) Torque reference and feedback. (**c**) Phase current.

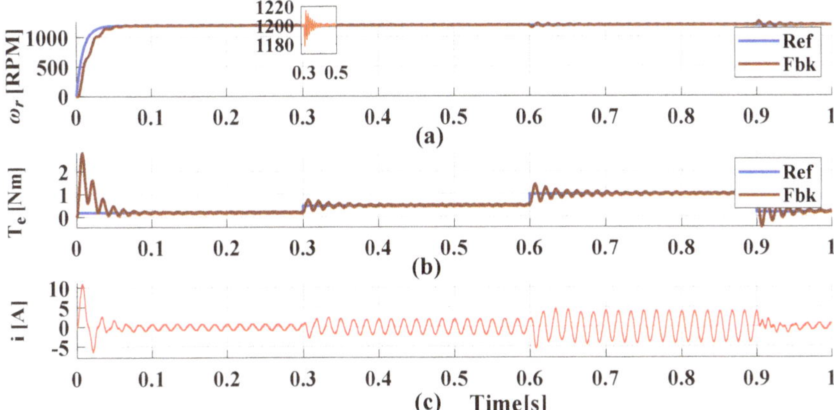

Figure 13. Simulation result of the 9-level inverter: (**a**) Speed reference and its feedback. (**b**) Torque reference and its feedback. (**c**) Phase current.

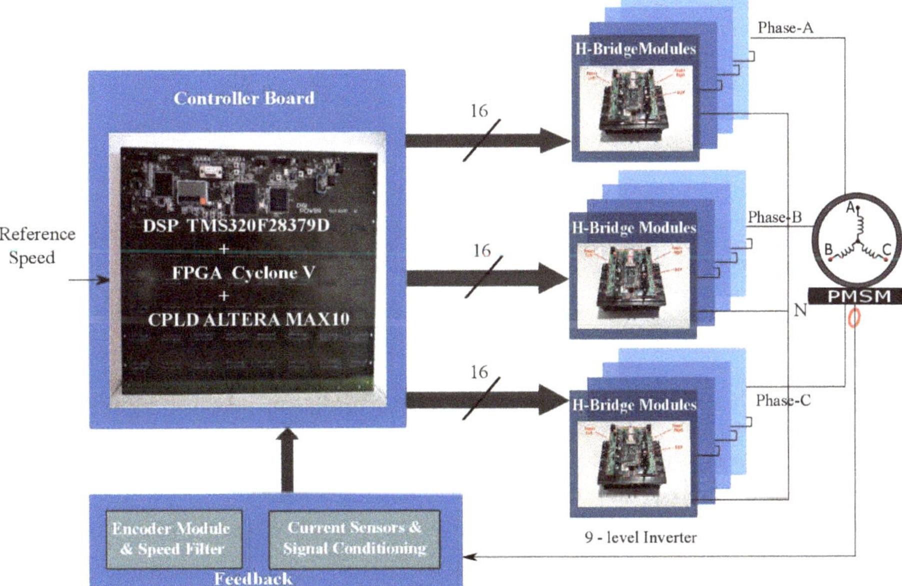

Figure 14. Block diagram of hardware implementation of the 9-level inverter drive.

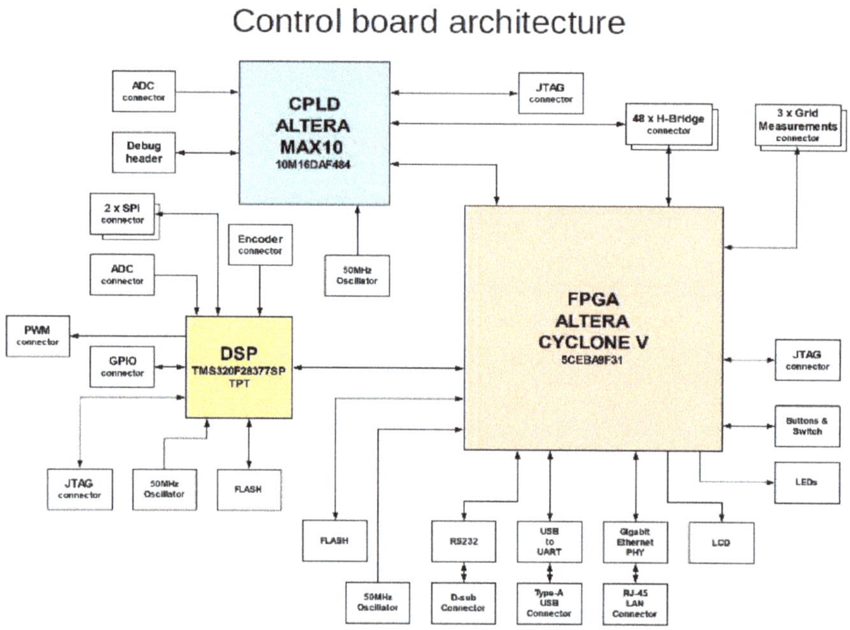

Figure 15. Control board architecture.

Control board is capable to control up to 48 H-bridges and offers large computational capability and a full set of I/O channels including Ethernet [23]. The complete test rig is shown in Figure 16. Each H-bridge has its own DSP and both voltage and current sensors; communications between the control board and H-bridges are bidirectional, therefore the whole system has intrinsic fault tolerance capability. Each H-bridge will provide local fault details, then if a fault occurs in power switches, the local DSP will communicate with the control board which can isolate that H-bridge and reconfigure the system, if the fault tolerant algorithm is implemented. The control logic block of FOC of the drive is implemented in DSP. The reference speed is the input of DSP, PI controllers are implemented in DSP, which generates the reference voltages V_A, V_B and V_C. These reference voltages are scaled down from 0 to 8 levels and PWM pulses are generated and sent to FPGA, where the switching states for each level are generated with dead time equal to 4 μs. The speed of the motor is sensed using incremental encoder, and it is calibrated in DSP. The currents of phases A and B are sensed using two LEM sensors, which are calibrated in DSP. The motor specifications are as shown in Table 2.

Table 2. Specifications of permanent magnet synchronous motor (PMSM).

Parameter	Symbols	Values
Rated power	P	6 [kW]
Rated current	I	5.3 [A]
Rated torque	T_e	3 [Nm]
Stator winding resistance	R_s	0.15 [Ω]
Stator winding inductance	L	0.85 [mH]
Rotor inertia	J	2.9 [Kg.m^2]
Pole pair	p	3
Rated speed	ω	6000 [RPM]
Permanent magnet flux	ϕ	0.0557 [Wb]

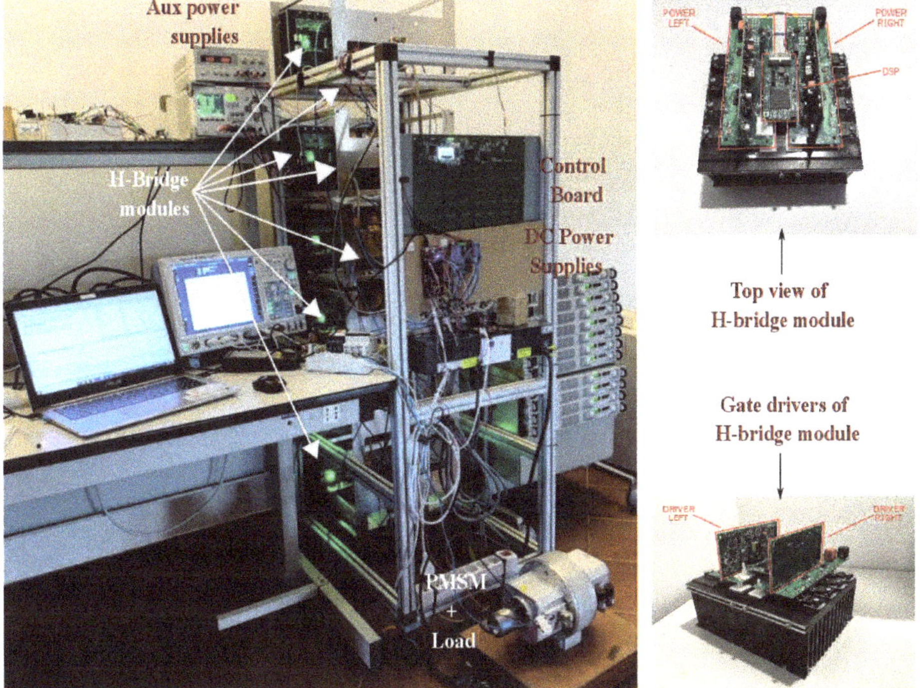

Figure 16. Test rig of cascaded H-bridge (CHB) 9-level inverter drive.

Figure 17a shows the experimental phase voltage and phase current waveform. Figure 17b shows the speed feedback when the motor is accelerated and decelerated. In particular, following the speed profile: from 3000 RPM to −3000 RPM to 3000 RPM to 600 RPM to −600 RPM to −3000 RPM, the speed feedback follows the speed reference, as shown in Figure 17b. It is also visible that when there is an increase in the speed reference, the phase current increases correspondingly. Figure 18a shows i_d and i_q when the motor is accelerated from 1000 RPM to 3600 RPM. This change affects mainly i_q, while i_d remains fairly constant. Similarly, when the speed is reduced from 3600 RPM to 1000 RPM Hz, the same behavior is observed as shown in Figure 18b. When the speed changes from −3000 RPM to 600 RPM, a little variation of torque is observed as shown in Figure 19a. A magnification of the last Figure is given in Figure 19b. The harmonic spectrum of the 9-level inverter is shown in the Figure 20. The span of the figure is 40 kHz, and the center is at 20 kHz. It is visible that the dominant harmonic present is at 33rd harmonic order, which is 10 kHz. This dominant harmonic is magnified and shown in the figure with its sidebands, validating the simulated harmonic spectrum.

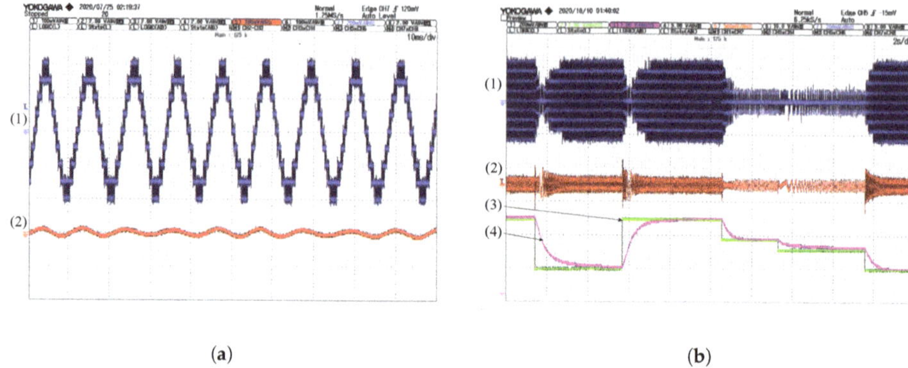

(a) (b)

Figure 17. Experimental result for the 9-level inverter: (**a**) (**1**) Phase voltage (y-axis: 20 V/div), (**2**) phase current (y-axis: 2 A/div). (**b**) (**1**) Phase voltage (y-axis: 20 V/div), (**2**) phase current (2 A/div), (**3**) speed reference (y-axis: 1820 RPM/div), (**4**) speed feedback (y-axis: 1820 RPM/div).

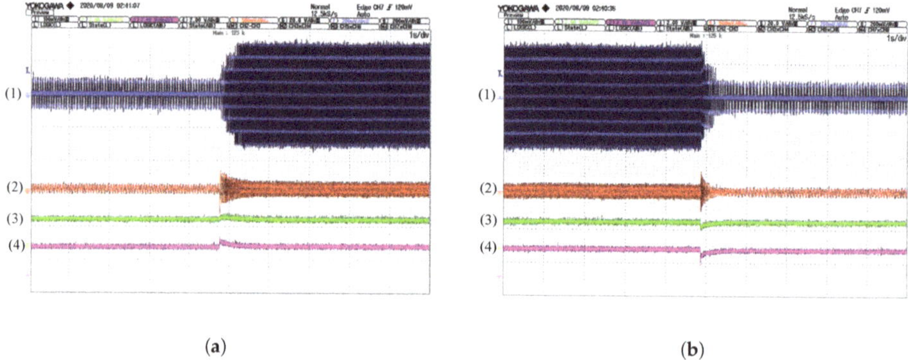

(a) (b)

Figure 18. Experimental result for the 9-level inverter: (**a**) (**1**) Phase voltage (y-axis: 20 V/div), (**2**) phase current (y-axis: 5 A/div), (**3**) i_d (y-axis: A/div), (**4**) i_q (y-axis: A/div). (**b**) (**1**) Phase voltage (y-axis: 20 V/div), (**2**) phase current (y-axis: 5 A/div), (**3**) i_d (y-axis: A/div), (**4**) i_q (y-axis: A/div).

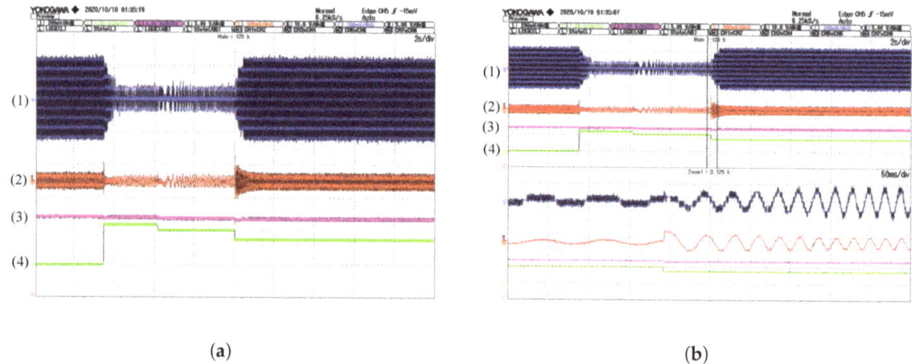

(a) (b)

Figure 19. Experimental result when speed is varied for the 9-level inverter: (**a**) (**1**) Phase voltage (y-axis: 20 V/div), (**2**) phase current (y-axis: 5 A/div), (**3**) torque (y-axis: Nms), (**4**) speed reference (y-axis: 3600 RPM/div). (**b**) (**1**) Phase voltage (y-axis: 20 V/div), (**2**) phase current (y-axis: 5 A/div), (**3**) torque (y-axis: Nm), (**4**) speed reference (y-axis: 3600 RPM/div).

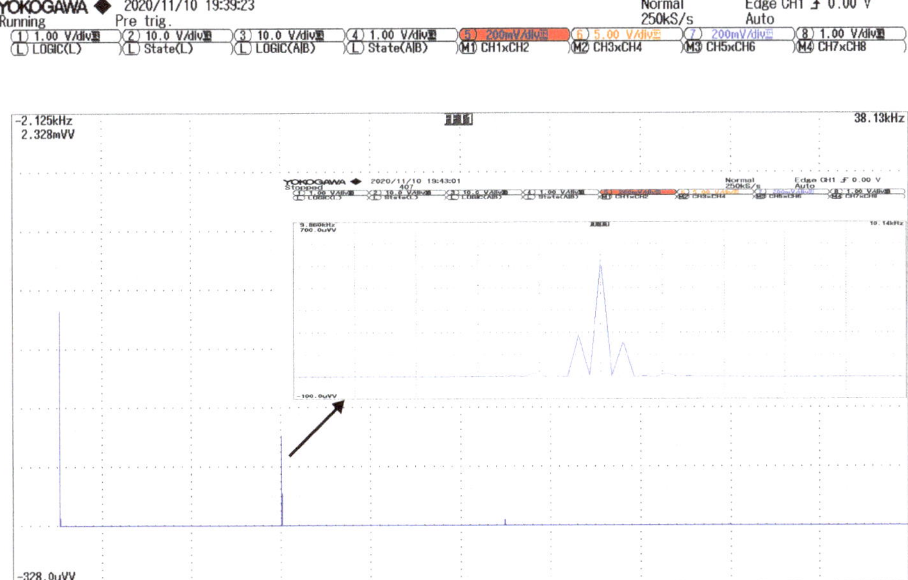

Figure 20. Experimental harmonic spectrum of phase voltage for the 9-level inverter.

5. Conclusions

The growth of FEA applications implies the development of large scale capability power converters. In this paper, a CHB 9-level inverter feeding PMSM for FEA applications has been proposed. It has been shown that the THD% reduces as the number of levels are increased, in turn reducing dv/dt. DC-link voltage is shared between the H-bridge modules as number of level increases, consequently it is possible to use low rated devices. The torque ripple is also reduced in comparison to TLI; the efficiency and the life of the motor are also improved. The adopted prototype of the multi-level converter developed by DigiPower can be used with high power motors and with multi-phase machines and in fault tolerant configurations. Experimental results have been presented to validate the theoretical analysis.

Author Contributions: The authors: V.P., C.B. and C.C. have equally contributed to this paper (conceptualization, software implementation and validation), moreover, they provided to formal analysis and preparation of the manuscript along the whole review process. C.C. is principal investigator of the founding project. All authors have read and agreed to the published version of the manuscript.

Funding: This research was funded by PRIN2017 OF Funder grant number MS9F49 G.

Acknowledgments: The authors would like to thank the DigiPower srl. and its staff members for the help in development of experimental setup.

Conflicts of Interest: The authors declare no conflict of interest.

Abbreviations

The following abbreviations are used in this manuscript:

PE	Power electronics
MLI	Multilevel inverter
FEA	Full electric aircraft
CHB	Cascaded H-bridge
THD	Total harmonic distortion

MI Modulation index
PMSM Permanent magnet synchronous motor
P I Proportional integral

References

1. Dorn-Gomba, L.; Ramoul, J.; Reimers, J.; Emadi, A. Power Electronic Converters in Electric Aircraft: Current Status, Challenges, and Emerging Technologies. *IEEE Trans. Transp. Electrif.* **2020**, *6*, 1648–1664.
2. Wheeler, P. Technology for the more and all electric aircraft of the future. In Proceedings of the 2016 IEEE International Conference on Automatica (ICA-ACCA), Curicó, Chile, 19–21 October 2016; pp. 1–5.
3. Rolandberger. Available online: https://www.rolandberger.com/en/Point-of-View/Electric-propulsion-is-finally-on-the-map.html (accessed on 1 October 2020).
4. AIRBUS. Available online: https://www.airbus.com/innovation/zero-emission/urban-air-mobility/cityairbus.html (accessed on 1 October 2020).
5. AIRBUS. Available online: https://www.airbus.com/innovation/zero-emission/electric-flight/e-fan-x.html (accessed on 1 October 2020).
6. Sarfan. Available online: https://www.safran-group.com/media/safran-proud-power-bell-nexus-20190107 (accessed on 1 October 2020).
7. Calzo, G.L.; Zanchetta, P.; Gerada, C.; Gaeta, A.; Crescimbini, F. Converter topologies comparison for more electric aircrafts high speed Starter/Generator application. In Proceedings of the 2015 IEEE Energy Conversion Congress and Exposition (ECCE), Montreal, QC, Canada, 20–24 September 2015; pp. 3659–3666.
8. Patel, V.; Buccella, C.; Saif, A.M.; Tinari, M.; Cecati, C. Performance Comparison of Multilevel Inverters for E-transportation. In Proceedings of the 2018 International Conference of Electrical and Electronic Technologies for Automotive, Milano, Italy, 9–11 July 2018; pp. 1–6.
9. Tolbert, L.M.; Peng, F.Z.; Habetler, T.G. Multilevel converters for large electric drives. *IEEE Trans. Ind. Appl.* **1999**, *35*, 36–44.
10. Rodriguez, J.; Lai, J.-S.; Peng, F.Z. Multilevel inverters: A survey of topologies, controls, and applications. *IEEE Trans. Ind. Electron.* **2002**, *49*, 724–738.
11. Yin, S.; Tseng, K.J.; Simanjorang, R.; Liu, Y.; Pou, J. A 50-kW High-Frequency and High-Efficiency SiC Voltage Source Inverter for More Electric Aircraft. *IEEE Trans. Ind. Electron.* **2017**, *64*, 9124–9134.
12. Zhang, D.; He, J.; Pan, D.; Dame, M.; Schutten, M. Development of Megawatt-Scale Medium-Voltage High Efficiency High Power Density Power Converters for Aircraft Hybrid-Electric Propulsion Systems. In Proceedings of the 2018 AIAA/IEEE Electric Aircraft Technologies Symposium (EATS), Cincinnati, OH, USA, 12–14 July 2018; pp. 1–5.
13. Viola, F. Experimental Evaluation of the Performance of a Three-Phase Five-Level Cascaded H-Bridge Inverter by Means FPGA-Based Control Board for Grid Connected Applications. *Energies* **2018**, *11*, 3298.
14. Glen Research Center. NASA. Available online: https://www1.grc.nasa.gov/aeronautics/eap/larger-aircraft/#converters (accessed on 1 October 2020).
15. Jack, A.G.; Mecrow, B.C.; Haylock, J.A. A comparative study of permanent magnet and switched reluctance motors for high-performance fault-tolerant applications. *IEEE Trans. Ind. Appl.* **1996**, *32*, 889–895.
16. Mitcham, A.J.; Cullen, J.J.A. Permanent magnet generator options for the More Electric Aircraft. In Proceedings of the International Conference on Power Electronics, Machines and Drives, Sante Fe, NM, USA, 4–7 June 2002; pp. 241–245.
17. Jia, Y.; Rajashekara, K. An Induction Generator-Based AC/DC Hybrid Electric Power Generation System for More Electric Aircraft. *IEEE Trans. Ind. Appl.* **2017**, *53*, 2485–2494.
18. Cao, W.; Mecrow, B.C.; Atkinson, G.J.; Bennett, J.W.; Atkinson, D.J. Overview of Electric Motor Technologies Used for More Electric Aircraft (MEA). *IEEE Trans. Ind. Electron.* **2012**, *59*, 3523–3531.
19. Patel, V.; Tinari, M.; Buccella, C.; Cecati, C. Analysis on Multilevel Inverter Powertrains for E-transportation. In Proceedings of the 2019 IEEE 13th International Conference on Compatibility, Power Electronics and Power Engineering (CPE-POWERENG), Sonderborg, Denmark, 23–25 April 2019; pp. 1–6.
20. Kanchan, R.S.; Baiju, M.R.; Mohapatra, K.K.; Ouseph, P.P.; Gopakumar, K. Space vector PWM signal generation for multilevel inverters using only the sampled amplitudes of reference phase voltages. *IEE Proc. Electr. Power Appl.* **2005**, *152*, 297–309.

21. Hemmati, R.; Wu, F.; El-Refaie, A. Survey of insulation systems in electrical machines. In Proceedings of the 2019 IEEE International Electric Machines & Drives Conference (IEMDC), San Diego, CA, USA, 12–15 May 2019; pp. 2069–2076.
22. Seri, P.; Montanari, G.C.; Hebner, R. Partial discharge phase and amplitude distribution and life of insulation systems fed with multilevel inverters. In Proceedings of the 2019 IEEE Transportation Electrification Conference and Expo (ITEC), Detroit, MI, USA, 19–21 June 2019; pp. 1–6.
23. Digipower Srl. Available online: https://www.digipower.it (accessed on 13 October 2020).

Article

Frequency-Domain Modeling of Harmonic Interactions in Voltage-Source Inverters with Closed-Loop Control

Malte John *,† **and Axel Mertens**

Institute for Drive Systems and Power Electronics, Leibniz University of Hannover, 30169 Hannover, Germany; mertens@ial.uni-hannover.de

* Correspondence: malte.john@volkswagen.de

† Current address: Volkswagen AG, 34219 Baunatal, Germany.

Received: 7 October 2020; Accepted: 5 November 2020; Published: 6 November 2020

Abstract: Power electronic converters, together with their loads, sources, and controls, form a coupled system that includes many nonlinear interactions, for instance due to pulse-width modulation (PWM) and feedback control. In this paper we develop a complete, nonlinear modeling approach for voltage-source inverters in the frequency domain, taking into account the harmonic components introduced into the system from the inputs and from the nonlinear digital PWM. The most important contribution is a method for analyzing how these harmonics propagate through the nonlinear system in steady state. To enable this, an analytic model of PWM with arbitrary, multiple-frequency input is necessary. A revised model of Asymmetrical regularly-sampled double-edge PWM (AD-PWM) is proposed and its incorporation into the system model regarding sampling effects is discussed. The resulting nonlinear equation system is numerically and simultaneously solved, yielding the spectra of all relevant signals in the converter. The results are validated with time-domain simulations and with measurements, proving the effectiveness of the proposed approach.

Keywords: harmonic analysis; power converters; pulse-width modulation (PWM); frequency-domain model; voltage-source inverter (VSI); closed-loop control

1. Introduction

The accurate assessment of current and voltage harmonics of power electronic systems is studied to meet a variety of design goals, including determination of dc-link capacitor size and lifetime [1,2], compliance with grid codes [3], avoidance of the excitation of resonances [4], and design of active power filters [5,6]. Power electronic systems, for example the voltage-source inverter (VSI) with its passive components and control (Figure 1), form closely coupled relationships that result in a nonlinear closed-loop system. The analysis of the algebraic equation system in the frequency domain is a complex process but supports a deeper understanding of the behavior of the system. By direct solution of the equation system, the system harmonics can be assessed.

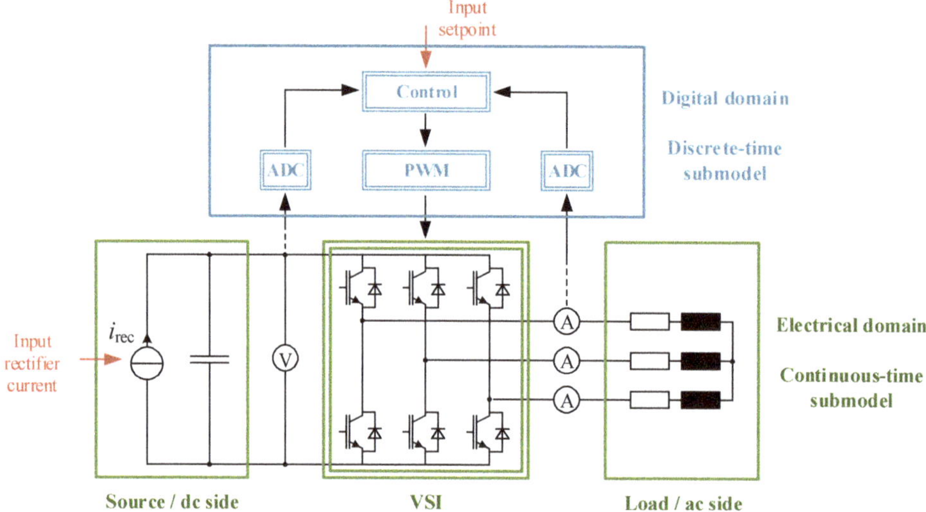

Figure 1. Topology of the example system. The two-level VSI connects a dc source to a three-phase passive load. The dc-link voltage and the ac-side currents are fed back via Analog-to-digital converters (ADCs) to the digital control, which regulates the switching state of the VSI via a pulse-width modulation (PWM). The power converter forms a closed-loop system, which is nonlinear due to the double framed blocks. The digital control in blue contains discrete-time signals and the electrical system in green is described by continuous-time signals. The system model has the rectifier current and the control setpoint as input signals displayed in red.

An important breakthrough in describing the frequency-domain behavior of switched power converters was the analysis of modulated signals. The double Fourier series expression was adapted by Bowes and Bird [7] to characterize the output spectrum of PWM converters. After further extension by Holmes [8], this method became a standard used to describe the switching spectrum of PWM VSI [9]. The double Fourier series analysis assumes in its original form:

1. A constant dc-link voltage, i.e., no interaction of dc side and ac side
2. A single frequency modulator input signal, i.e., no interaction with the control loop
3. Ideal switching of the power electronics, i.e., no dead time and lossless switches

To calculate the ac-side voltage spectrum under the influence of a variable dc-link voltage, a convolution of the switching function spectrum and the dc-link voltage spectrum can be performed. As shown by McGrath et al. [2], the dc-link current spectrum can be calculated in a similar way by using a convolution of the switching function spectrum and the ac-side current spectrum.

Subsequently, modifications to the double Fourier method as well as alternative methods evolved, characterizing the PWM output spectrum [10]. Special emphasis was given to incorporating modulator input signals that contain multiple frequency components [5,11–13] and dc-link voltage oscillations [14,15]. Because the interaction of the signals within the power stage of the plant and the influence of the closed-loop control system are not examined, the research question arises of how to integrate these interactions.

A common approach for modeling and analyzing PWM converters is the averaging method, introduced by Wester, Ćuk, and Middlebrook [16–18] and extended by Erickson and Maksimovic [19–21] and Hiti [22]. The averaging method is useful in developing linearized small-signal models and is commonly implemented in impedance-based models and stability analysis [23]. Corradini et al. [24] utilized this method in describing the digital control of power converters, introducing small-signal delays to consider sampling effects of the PWM and the digital control. Almér and Jönsson [25] developed a dynamic phasor model of a dc-dc converter with closed-loop control applying averaging and truncation of the high-frequency components. By applying small-signal assumptions, Yue et al. [26] modeled frequency components that are unsynchronized with and close to the sampling frequency and cause undesirable beat-frequency components due to aliasing.

The averaging and linearization methods truncate the high frequency components caused by the modulation and switching process. Modeling errors can result when the switching-band and the base-band components overlap and sampling effects like aliasing occur. A need for research was identified to develop a frequency-domain method that models a power electronic system including its harmonic interactions. These interactions concern the mutual dependencies between the continuous signals of the power stage, the digital signals of the control, and the influence of nonlinear effects such as modulation, sampling, and aliasing.

This article deals with the prediction of the harmonic spectrum in a converter system in the frequency domain. A special focus arises from the description of the interactions that arise in the closed control loop. We want to convey a deeper understanding of the harmonic interactions and as a result present a modeling method that enables the calculation of the spectra including their nonlinear interactions. Some modeling aspects are well known (converter model, linear components, ADC, and PI control) and they are included here for completeness of the model. The complexity and novelty of this paper lies in the PWM model's correct consideration of the sampling process and the interactions of the subsystems. This stands in contrast to the prevailing models found in the presented literature, which either model the components with their input-to-output behavior neglecting the closed-loop interactions or with a linearized behavior.

The control is developed for the example of a VSI with closed-loop current control. The influence of quantization is excluded in the models. The approach to describe the spectrum of PWM published by Song and Sarwate [11] is utilized. A review of their results for AD-PWM revealed deviations from measurement results. Therefore, we present a revised derivation of the switching spectrum for AD-PWM. The method applies for hard-switched converters using a fixed switching frequency in the linear modulation range. Furthermore, the incorporation of the PWM model into the total system model as a hybrid of discrete-time and continuous-time signals is presented with special emphasis on the consideration of sampling effects. The resulting models are evaluated with band-limited Fourier coefficients for an example case in steady state and compared with time-domain simulations and experimental results. The work presented here is a part of the author's dissertation [27].

2. Models of Individual Components and Effects

The example system used in this paper (Figure 1) consists of a dc source, a three-phase two-level VSI, and an RL-load. The closed-loop control of the load currents provide the ac-side voltage setpoints for the VSI, which are divided by half the measured dc-link voltage and transformed into gate signals by a pulse-width modulation (PWM) process. The influence of the dc source (e.g., a rectifier) is modeled by a current source i_{rec}. This enables the incorporation of the harmonic behavior of a rectifier but neglects the influence of the dc-link voltage on the rectifier.

A digital control system is assumed, because of its widespread application and to discuss the effects on the modeling approach with both discrete-time and continuous-time signals being present in the system. Analog sensors measure the required voltage and current values, which are then sampled and held by an ADC. In order to meet the control goal determined by the input setpoint signals, the control outputs duty cycles that are proportional to the required ac-side voltages. Applying a PWM to the duty cycles generates the gate signals, which control the power semiconductors. In the following sections the models of the individual components of the system are introduced.

2.1. Switching Power Converter

The VSI in Figure 1 consists of three half bridges. Assuming ideal switching behavior (loss-less switching and no dead time), each half bridge of phase $\nu \in \{1, 2, 3\}$ is represented by a single-pole double-throw switch with the switching function s_ν, resulting in the equivalent circuit of the electrical subsystem in Figure 2. The switching function is used to provide the algebraic connection between the ac-side voltages $u_{\mathrm{ac}\nu}$ and the dc-side voltage u_{dc}, with

$$u_{\mathrm{ac}\nu}(t) = \frac{1}{2} \cdot s_\nu(t) \cdot u_{\mathrm{dc}}(t), \tag{1}$$

and between the ac-side currents $i_{\mathrm{ac}\nu}$ and the dc-side current i_{dc}, with

$$i_{\mathrm{dc}}(t) = \sum_{\nu=1}^{3} \frac{1}{2} \cdot s_\nu(t) \cdot i_{\mathrm{ac}\nu}(t). \tag{2}$$

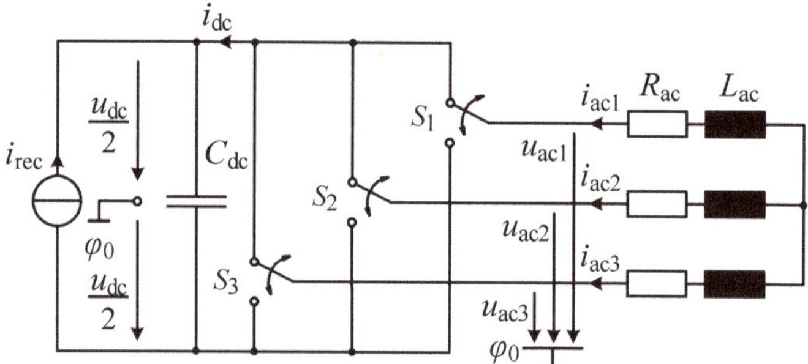

Figure 2. Equivalent circuit of the three-phase two-level VSI.

Transforming the equations to the frequency domain yields a convolution of the spectra, with

$$U_{\mathrm{ac}\nu}(f) = \frac{1}{2} \cdot S_\nu(f) * U_{\mathrm{dc}}(f), \tag{3}$$

$$I_{\mathrm{dc}}(f) = \sum_{\nu=1}^{3} \frac{1}{2} \cdot S_\nu(f) * I_{\mathrm{ac}\nu}(f), \tag{4}$$

where $*$ denotes the convolution operator.

2.2. Load and Source Models

The load and source models provide the mathematical model for the connection between the voltages and currents on either side of the converter. In the common modeling case of linear passive components, the ac-side impedance representing a grid filter, a motor stray inductance or a passive load, and the dc-link capacitor are modeled as linear devices. For a symmetrical ac-side impedance, the frequency-domain expression of the ac side is

$$\begin{bmatrix} I_{\mathrm{ac1}}(f) \\ I_{\mathrm{ac2}}(f) \\ I_{\mathrm{ac3}}(f) \end{bmatrix} = -\frac{1}{3} \cdot (R_{\mathrm{ac}} + \mathrm{j}2\pi f L_{\mathrm{ac}}) \cdot \begin{bmatrix} 2 & -1 & -1 \\ -1 & 2 & -1 \\ -1 & -1 & 2 \end{bmatrix} \cdot \begin{bmatrix} U_{\mathrm{ac1}}(f) \\ U_{\mathrm{ac2}}(f) \\ U_{\mathrm{ac3}}(f) \end{bmatrix}, \tag{5}$$

where R_{ac} is the ac-side resistance and L_{ac} is the ac-side inductance.

The voltage-stiff dc link of a VSI consists of a large capacitance C_{dc}. Its frequency-domain representation results in

$$U_{\mathrm{dc}}(f) = \frac{1}{2\pi f C_{\mathrm{dc}}} \cdot (I_{\mathrm{dc}}(f) + I_{\mathrm{rec}}(f)), \tag{6}$$

Please note that without a further resistive component in the dc-link, (6) contains a singularity, which results in a degree of freedom for $U_{\mathrm{dc}}(0)$ in the model.

2.3. Analog-to-Digital Conversion and Filtering

In order to provide feedback of the currents and voltages to the control system, the signals are sensed using analog sensors and then converted into digital signals. The block diagram of the measurement system that is used to track a continuous-time signal is depicted in Figure 3. The spectrum of the signal $X(f)$ represents I_{acv} or U_{dc}.

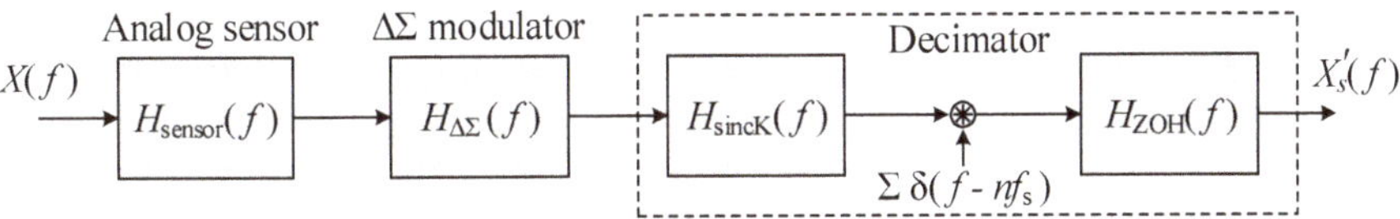

Figure 3. Structure of the measurement system.

The analog sensors are represented by a first-order low-pass filter with a cut-off frequency f_{cut},

$$H_{sensor}(f) = \frac{1}{j2\pi f / f_{cut} + 1}. \tag{7}$$

In this paper, a $\Delta\Sigma$ converter is used as ADC, which is assumed to have a sufficiently high oversampling ratio. Therefore, its influence is low and $H_{\Delta\Sigma}(f) \approx 1$.

A decimator is used to reduce the quantization noise of the sampled signal. The decimator has the additional tasks of preventing aliasing and downsampling to the controller frequency while the signal's word size is increased. The decimation is often performed by sinc filters of order $K > 1$ [28], which are described by a transfer function of

$$H_{sincK}(f) = \left(\frac{f_s}{f_{\Delta\Sigma}} \cdot \frac{1 - e^{-j2\pi f / f_s}}{1 - e^{-j2\pi f / f_{\Delta\Sigma}}} \right)^K, \tag{8}$$

where $f_{\Delta\Sigma}$ is the sampling frequency of the $\Delta\Sigma$ modulator.

The output of the digital filter is sampled with the sampling frequency of the control f_s. The sampling process is represented by a multiplication of the continuous signal $x(t)$ with a sequence of Dirac pulses δ, which results in a continuous-time representation of a sampled signal $x_s(t)$, with

$$x_s(t) = x(t) \cdot \sum_{n=-\infty}^{\infty} \delta(t - n/f_s). \tag{9}$$

Transformation of (9) into the Fourier domain results in

$$X_s(f) = f_s \cdot X(f) * \sum_{n=-\infty}^{\infty} \delta(f - nf_s), \tag{10}$$

where $X_s(f)$ is known as the Discrete-time Fourier transform (DTFT), which differs from the Continuous-time Fourier transform (CTFT) $X(f)$. This model of the sampler enables the consideration of sampling effects such as aliasing. Moreover, the sampling process is commonly synchronized with the PWM period, minimizing the influence of switching harmonics in the sensed signals [29]. This effect is included in this modeling approach and a time-shift between the PWM signal and the sampling points can easily be considered in (9) and (10).

Finally, the sampled signals are held in the control system for one control period, which is modeled with a zero-order hold (ZOH) block, with

$$H_{ZOH}(f) = e^{-j\pi f T_s} \cdot \mathrm{si}(\pi f T_s). \tag{11}$$

The spectrum of the resulting sampled and low-pass filtered signal is written with an apostrophe $X'_s(f)$ to distinguish it from the original spectrum $X(f)$.

2.4. Control System

In this paper PI current control in the rotating reference frame is used as an example control system. The control system is shown in Figure 4. The inputs are the sampled ac-side currents $i'_{s,ac}$ and dc-link voltage $u'_{s,dc}$ and the current setpoints $i_{s,sp}$. The outputs of the current control are the setpoint voltages in 123 coordinates $u_{s,sp}$, which function as inputs of the duty-cycle calculation. The outputs of the complete control block are the duty-cycle commands for the PWM process d_s.

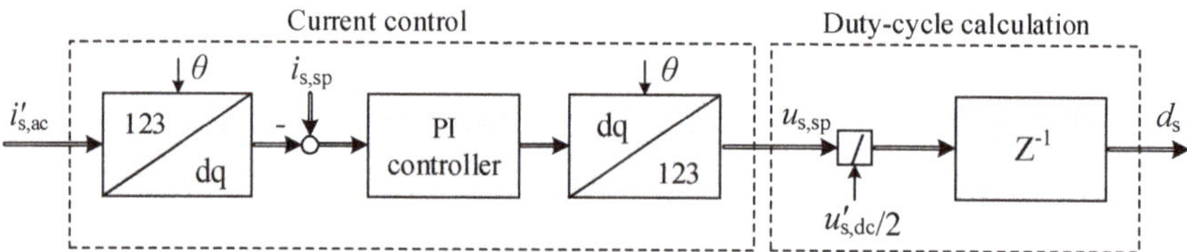

Figure 4. Block diagram for closed-loop control using PI current control and dc-link voltage feedback.

2.4.1. Current Control

The current control is implemented in the rotating reference frame by applying the Park transformation to the input signals. The transformation consists of the Clarke transformation and a rotation matrix. The measured ac-side current spectra are first transformed into current spectra in αβ coordinates, with

$$\begin{bmatrix} I_{s,\alpha}(f) \\ I_{s,\beta}(f) \end{bmatrix} = \frac{2}{3} \cdot \begin{bmatrix} 1 & -\frac{1}{2} & -\frac{1}{2} \\ 0 & \frac{\sqrt{3}}{2} & -\frac{\sqrt{3}}{2} \end{bmatrix} \cdot \begin{bmatrix} I'_{s,ac1}(f) \\ I'_{s,ac2}(f) \\ I'_{s,ac3}(f) \end{bmatrix}. \tag{12}$$

Then a rotation into the dq reference frame is performed, which leads to a frequency shift in the current spectra [30]. For this it is assumed that the rotating angle $\theta(t)$ has a constant gradient with $\theta(t) = 2\pi f_0 t$, where f_0 is the fundamental frequency. The current spectra in dq coordinates result in

$$\begin{bmatrix} I_{s,d}(f) \\ I_{s,q}(f) \end{bmatrix} = \frac{1}{2} \cdot \begin{bmatrix} 1 & -j \\ j & 1 \end{bmatrix} \cdot \begin{bmatrix} I_{s,\alpha}(f-f_0) \\ I_{s,\beta}(f-f_0) \end{bmatrix} + \frac{1}{2} \cdot \begin{bmatrix} 1 & j \\ -j & 1 \end{bmatrix} \cdot \begin{bmatrix} I_{s,\alpha}(f+f_0) \\ I_{s,\beta}(f+f_0) \end{bmatrix}. \tag{13}$$

The current controller determines the setpoint voltages of the VSI in dq coordinates under the influence of the setpoint signals $I_{s,sp,d}$, $I_{s,sp,q}$, with

$$\begin{bmatrix} U_{s,sp,d} \\ U_{s,sp,q} \end{bmatrix} = H_{PI}(f) \cdot \begin{bmatrix} I_{s,sp,d}(f) \\ I_{s,sp,q}(f) \end{bmatrix} - \begin{bmatrix} H_{PI}(f) & -2\pi f_0 L_{ac} \\ 2\pi f_0 L_{ac} & H_{PI}(f) \end{bmatrix} \cdot \begin{bmatrix} I_{s,d}(f) \\ I_{s,q}(f) \end{bmatrix}. \tag{14}$$

A disturbance compensation of the cross-coupling of the dq axes is taken into account on the counterdiagonal of the matrix. The linear PI controllers are described by a transfer function

$$H_{PI}(f) = K_P + \frac{K_P}{T_I} \cdot H_I(f), \tag{15}$$

where K_P is the proportional gain and T_I is the integrator time constant. In the case of backward Euler, the transfer function of the integrator $H_I(f)$ is described by

$$H_I(f) = T_s \cdot \frac{e^{j2\pi f T_s}}{e^{j2\pi f T_s} - 1}. \tag{16}$$

The dc component of the controller's transfer function has infinite gain, with $H_{\mathrm{I}}(f \rightarrow 0) \rightarrow \infty$ and the dc components of the sampled dq currents equals the setpoint values. Consequently, the dc components of the setpoint voltages $U_{\mathrm{s,sp,d}}(0)$ and $U_{\mathrm{s,sp,q}}(0)$ cannot be calculated by (14). Their solutions result instead from the complete equation system that incorporates the interaction of the control with the physical model of the VSI and its load.

Finally, the setpoint voltages are transformed back into 123 coordinates, with

$$\begin{bmatrix} U_{\mathrm{s,sp},\alpha}(f) \\ U_{\mathrm{s,sp},\beta}(f) \end{bmatrix} = \frac{1}{2} \cdot \begin{bmatrix} 1 & \mathrm{j} \\ -\mathrm{j} & 1 \end{bmatrix} \cdot \begin{bmatrix} U_{\mathrm{s,sp,d}}(f - f_0) \\ U_{\mathrm{s,sp,q}}(f - f_0) \end{bmatrix} + \frac{1}{2} \cdot \begin{bmatrix} 1 & -\mathrm{j} \\ \mathrm{j} & 1 \end{bmatrix} \cdot \begin{bmatrix} U_{\mathrm{s,sp,d}}(f + f_0) \\ U_{\mathrm{s,sp,q}}(f + f_0) \end{bmatrix}, \tag{17}$$

$$\begin{bmatrix} U_{\mathrm{s,sp,1}}(f) \\ U_{\mathrm{s,sp,2}}(f) \\ U_{\mathrm{s,sp,3}}(f) \end{bmatrix} = \frac{2}{3} \cdot \begin{bmatrix} 1 & 0 \\ -\frac{1}{2} & \frac{\sqrt{3}}{2} \\ -\frac{1}{2} & -\frac{\sqrt{3}}{2} \end{bmatrix} \cdot \begin{bmatrix} U_{\mathrm{s,sp},\alpha}(f) \\ U_{\mathrm{s,sp},\beta}(f) \end{bmatrix}. \tag{18}$$

2.4.2. Duty-Cycle Calculation

The duty cycle is calculated by dividing the setpoint voltage of phase ν by half the measured dc-link voltage. The duty cycle that was calculated in sampling step k is committed to the modulator within the control cycle and is applied to the switching pattern in the next control step $k+1$, with

$$d_{\nu}[k+1] = \frac{u_{\mathrm{s,sp},\nu}[k]}{u'_{\mathrm{s,dc}}[k]/2}. \tag{19}$$

The inverse convolution in the frequency domain is used to achieve the equivalent operation of the division in the time domain. Because the dc-link voltage is the state variable of an energy storage it is required that $u_{\mathrm{dc}}(t) > 0, \forall t$ and $1/u_{\mathrm{dc}}(t)$ is a slowly growing signal. This means that the requirements for the existence of the inverse convolution are fulfilled.

Lastly, the computational delay is taken into account by the delay block (z^{-1}) in Figure 4. The calculation of the duty cycle thus results in

$$D_{\mathrm{s},\nu}(f) = 2 \cdot H_{\mathrm{delay}}(f) \cdot U'^{-1}_{\mathrm{s,dc}}(f) * U_{\mathrm{s,sp},\nu}(f), \tag{20}$$

where the delay is considered with

$$H_{\mathrm{delay}}(f) = \mathrm{e}^{-\mathrm{j}2\pi f T_{\mathrm{c}}}. \tag{21}$$

2.5. Model for PWM with Harmonic Input

The switching function is determined in the PWM process. In order to include the PWM process in a frequency-domain model that considers the interaction between the control and the electrical system, an expression of the switching function spectrum

$$S(f) = g\,(D(f)), \tag{22}$$

as an explicit nonlinear function g of the duty-cycle spectrum is needed. This can be found in the model by Song and Sarwate [11], which originates from a time-domain representation of the pulse pattern using the Heaviside function and is further developed into a frequency-domain representation. A model for naturally-sampled double-edge PWM (ND-PWM) is derived by Song and Sarwate with

$$
\begin{aligned}
S_{\mathrm{ND}}(f, D) = D(f) + S_{\mathrm{c}}(f) \cdot \mathrm{e}^{-\mathrm{j}\pi f T_{\mathrm{sw}}/2} + \sum_{m=1}^{\infty} (-1)^m \cdot \sum_{n=1}^{\infty} \ldots \\
\left(\frac{(\mathrm{j}2m\pi)^{2n-2}}{2^{2n-2} \cdot (2n-1)!} \cdot \left(D^{*(2n-1)}(f + 2mf_{\mathrm{sw}}) + D^{*(2n-1)}(f - 2mf_{\mathrm{sw}}) \right) \ldots \right. \\
\left. - \frac{(\mathrm{j}(2m-1)\pi)^{2n-1}}{\mathrm{j}2^{2n-1} \cdot (2n)!} \cdot \left(D^{*(2n)}(f + (2m-1)f_{\mathrm{sw}}) + D^{*(2n)}(f - (2m-1)f_{\mathrm{sw}}) \right) \right)
\end{aligned}
\tag{23}
$$

where

$$
S_{\mathrm{c}}(f) = \sum_{p=0}^{\infty} \frac{2}{\mathrm{j}\pi(2p+1)} \cdot \left[\delta(f - (2p+1)f_{\mathrm{sw}}) - \delta(f + (2p+1)f_{\mathrm{sw}}) \right] \quad , \tag{24}
$$

and $D^{*n}(f)$ denotes the Fourier transform of the duty cycle signal raised to the power of n:

$$
(d(t))^n \circ\!\!-\!\!\bullet\, D^{*n}(f). \tag{25}
$$

AD-PWM is characterized by a low total harmonic distortion level, compared to other discrete type PWM methods [9]. Due to its discrete-time property, it is easier to implement this modulation method on digital control systems in contrast to natural sampling [11]. A review of results for AD-PWM presented by Song and Sarwate ([11], (56)) revealed differences in comparison to simulation results from a validated time-domain model. A revised derivation of the analytical model for AD-PWM results in the switching function spectrum of

$$
S_{\mathrm{AD}}(f, D) = \mathrm{e}^{-\mathrm{j}\pi f T_{\mathrm{sw}}/2} \cdot \left[S_{\mathrm{c}}(f) + \sum_{m=-\infty}^{\infty} \sum_{n=1}^{\infty} \frac{(\mathrm{j}\pi f T_{\mathrm{sw}})^{n-1}}{2^n \cdot n!} \cdot D^{*n}(f - mf_{\mathrm{sw}}) \cdot \left(1 - (-1)^{m+n}\right) \right]. \tag{26}
$$

For AD-PWM, the duty cycle of the PWM is adjusted at two sampling points per PWM cycle. Hence, the duty cycles of the two halves of a pulse differ from each other and the pulse appears asymmetrically spaced from the center of the PWM period. In comparison to Song and Sarwate ([11], (56)), the revised equation includes the representation of the two sampling points. The derivation of (26) is presented in the Appendix A. A numerical evaluation of (26) follows in Section 4.2 with a comparison with a time-domain simulation and measurement results.

The internal structure of regularly-sampled PWM can be divided into a sampler, a ZOH, and a representation of naturally-sampled PWM. From this, two compositions of the frequency-domain model describing regularly-sampled PWM result, as demonstrated in Figure 5. Subfigure (a) represents the model described by (26), in which the input is the Fourier transform of the continuous-time duty cycle $D(f)$. The description of the sampler and the ZOH can be separated from the PWM model without loss of generality, using the Fourier transform of the discrete-time duty cycle $D_{\mathrm{s}}(f)$ as the input of a model of naturally-sampled double-edge PWM. This method is utilized to calculate the switching function spectrum depicted in subfigure (b). Therefore, regularly-sampled PWM is equivalent to naturally-sampled PWM when applying a "distorted" duty cycle spectrum $D_{\mathrm{s}}(f)$.

When considering feedback control, the digital controller outputs a discrete-time duty-cycle with a spectrum $D_{\mathrm{s}}(f)$ (Figure 4), which interacts with the signals of the PWM and the plant. As stated earlier, the regularly-sampled PWM model (Figure 5a) requires an input spectrum $D(f)$ of the continuous-time duty-cycle $d(t)$. However, as demonstrated in Figure 6a, a naturally-sampled PWM model may be applied to express regularly-sampled PWM under the condition that the input is the spectrum of the discrete-time signal $D_{\mathrm{s}}(f)$. Because the DTFT of the input is not band-limited, the consideration of a large spectrum is required to achieve high quality results using this model.

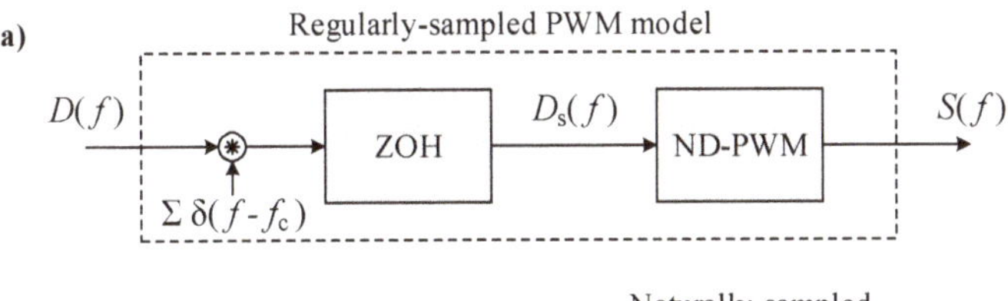

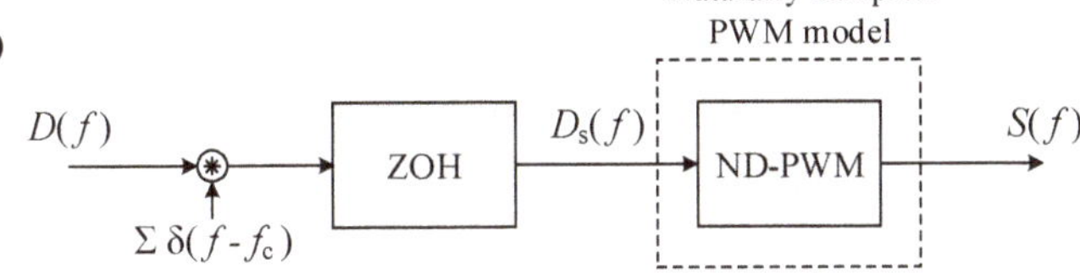

Figure 5. Equivalent methods representing regularly-sampled PWM when the input is the CTFT of the duty cycle $D(f)$: (**a**) The series connection of sampler, ZOH, and naturally-sampled PWM is combined in one mathematical expression. (**b**) The CTFT of the duty cycle $D(f)$ is first transferred into the DTFT of the duty cycle $D_s(f)$ using a model of the sampler and the ZOH. The resulting spectrum is used as the input for the naturally-sampled PWM model [27].

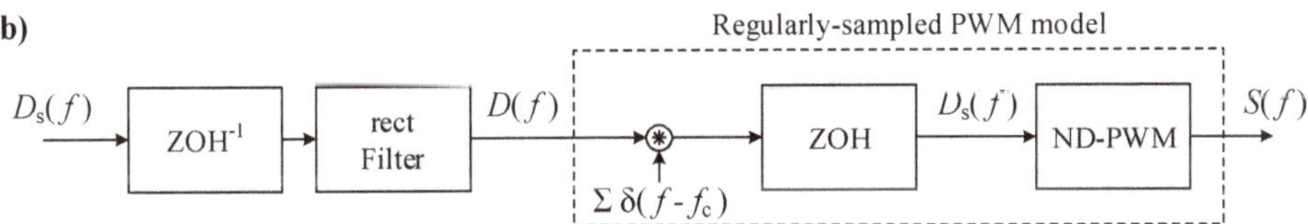

Figure 6. Equivalent methods representing regularly-sampled PWM when the input is the DTFT of the duty cycle $D_s(f)$: (**a**) A block expressing naturally-sampled PWM. (**b**) An inverse ZOH and a rectangular filter the ZOH block and sampler of a regularly-sampled PWM model [27].

An alternative method is presented in Figure 6b: The DTFT of the duty-cycle is converted into a CTFT, which is subsequently applied to the AD-PWM model. The benefit of this method is that input of the PWM model (dashed box in Figure 6b) is a band-limited signal and thus the consideration of a smaller spectrum is required to achieve the same quality as the method in Figure 6a. For further clarification, a numerical comparison is provided in this section. The conversion is performed by application of the inverse transfer function of the ZOH

$$\begin{aligned} H_{\mathrm{ZOH}}^{-1}(f) &= \frac{\mathrm{j}2\pi f T_{\mathrm{c}}}{1-\mathrm{e}^{-\mathrm{j}2\pi f T_{\mathrm{c}}}}, \quad f \neq 0 \\ H_{\mathrm{ZOH}}^{-1}(0) &= 1 \end{aligned} \tag{27}$$

and the inversion of the sampling process. Since $D_s(f)$ is the DTFT of a band-limited signal, the sampling process is inverted by a rectangular filter with

$$H_{\mathrm{rect}}(f) = \begin{cases} 1 & \text{for} \quad -f_{\mathrm{c}}/2 \leq f \leq f_{\mathrm{c}}/2 \\ 0 & \text{otherwise} \end{cases}. \tag{28}$$

A numerical comparison of the models in Figure 6a,b follows in Section 4.3.

3. System Model for Interactions

All equations needed to describe the overall converter system form an algebraic nonlinear equation system

$$\vec{g}(\vec{X}(f), \vec{U}(f)) = \vec{0}, \tag{29}$$

where $\vec{X}(f)$ are the spectra of the system's signals and $\vec{U}(f)$ are the spectra of the inputs.

As a simplification, the total equation system was numerically evaluated for the steady state. This enables the representation of the periodic time-domain signals $x(t)$ by a complex Fourier series. Furthermore, the Fourier series was band-limited to a maximum order of k_{max} so that a signal $x(t)$ was approximated by

$$x_1(t) \approx \sum_{k=-k_{\mathrm{max}}}^{k_{\mathrm{max}}} X_{1,k} \cdot \mathrm{e}^{\,\mathrm{j}k2\pi f_0 t}, \quad k \in \mathbb{Z}, \tag{30}$$

where $X_{1,k} \in \mathbb{C}$ is the complex Fourier coefficient of order k that represents the k-th harmonic of the fundamental frequency f_0.

The coefficients of the band-limited Fourier series form a discrete and finite spectrum that can be collected in a vector representation

$$\vec{X}_1 = [X_{1,-k_{\mathrm{max}}}, .., X_{1,-1}, X_{1,0}, X_{1,1}, .., X_{1,k_{\mathrm{max}}}]^{\mathrm{T}}. \tag{31}$$

Therefore, all input signals and system signals are represented by a vector of the length $l = 2\,k_{\mathrm{max}} + 1$.

The calculation of all spectra requires the simultaneous solution of the nonlinear equation system. The equation system chosen for the following numerical evaluation describes $N = 13$ signals, collected in a vector

$$\vec{X} = [\vec{I}_{\mathrm{ac1}}, \vec{I}_{\mathrm{ac2}}, \vec{I}_{\mathrm{ac3}}, \vec{U}_{\mathrm{dc}}, \vec{D}_{\mathrm{s,1}}, \vec{D}_{\mathrm{s,2}}, \vec{D}_{\mathrm{s,3}}, \vec{I}_{\mathrm{s,d}}, \vec{I}_{\mathrm{s,q}}, \vec{U}_{\mathrm{s,sp,\alpha}}, \vec{U}_{\mathrm{s,sp,\beta}}, \vec{U}_{\mathrm{s,sp,d}}, \vec{U}_{\mathrm{s,sp,q}}]^{\mathrm{T}}. \tag{32}$$

The size of the vector is $N \cdot l$, which are given by the input signals. All other signals are described as dependent variables and are not written explicitly in the equation system. However, it is possible to calculate them directly using the signals present in $\vec{X}$. The input signals of the system are the dc-side disturbance current, the dc component of the dc-link voltage, and the current setpoint in dq coordinates. Thus the input vector $\vec{U}$ contains the spectra of the ac-current setpoints $\vec{I}_{\mathrm{s,sp,d}}$, $\vec{I}_{\mathrm{s,sp,q}}$, the rectifier current $\vec{I}_{\mathrm{rec}}$, and the dc component of the dc-link voltage $\vec{U}_{\mathrm{dc}}(0)$.

Further approximations are necessary to enable a numerical evaluation:

- The infinite summation in (26) is limited to $-m_{\mathrm{max}} \leq m \leq m_{\mathrm{max}}$ and $-n_{\mathrm{max}} \leq n \leq n_{\mathrm{max}}$.
- The convolution of two vectors can be represented by a multiplication of the convolution matrix of one of the vectors with the second vector. The inverse convolution in (20) is calculated by a multiplication of $\vec{U}_{\mathrm{s,sp},\nu}$ with the inverse of the convolution matrix $\mathbf{C}(.)$ of $\vec{U}'_{\mathrm{dc}}$. The complete convolution matrix has the size $(2l-1) \times l$. To enable the inversion of the matrix, a truncated form of the convolution matrix is used that is square with $l \times l$. This form corresponds to the *Matlab* function *conv* when applying the option *same*. The truncated convolution matrix of a vector $\vec{X}$ is defined as

$$C(\vec{X}) = \begin{bmatrix} X_0 & \dots & X_{-k_{\max}} & & & & \\ X_1 & \dots & X_{(1-k_{\max})} & X_{-k_{\max}} & & & \\ \vdots & & \vdots & \vdots & & & \\ X_{k_{\max}} & & & \dots & & & X_{-k_{\max}} \\ & & & \vdots & \vdots & & \vdots \\ & & & X_{k_{\max}} & X_{(k_{\max}-1)} & \dots & X_{-1} \\ & & & & X_{k_{\max}} & \dots & X_0 \end{bmatrix}. \tag{33}$$

The numerical solution was executed in *Matlab* by applying the *trust-region algorithm* of the *Matlab* function *fsolve*. This algorithm was selected since it allows for the incorporation of the Jacobian matrix and a sparsity pattern. Unlike the complicated analytical derivation of the Jacobian, the analysis of a dependency between two variables is straightforward. This enables the formulation of a sparsity matrix, which indicates all partial derivatives of the input variables unequal to zero. Knowledge of the sparsity matrix enables the solver to accelerate the solution process by calculating only nonzero derivatives. In this particular case, less than 15% of the matrix elements were nonzero. By applying the sparsity pattern, the number of function calls for the numerical calculation of the Jacobian was halved.

A vector of initial values $\vec{X}_0$ is required for the iterative solution process, which is provided by a simple fundamental frequency model. It calculates the fundamental frequency components for ac quantities and the dc component of dc quantities, with all remaining components set to zero.

4. Results

4.1. Experimental System

The left photo in Figure 7 shows the interior part of the experimental system, where the IGBT modules (IFS150V12PT4) and the dc link are visible in the center. One IGBT module was used in the experiments as a VSI and the second module was switched off and used as a diode bridge rectifier.

The VSI control was implemented on a system-on-chip unit *Xilinx Zynq 7000*. It contains an FPGA and an *ARM Cortex-A9* double-core processor. The FPGA comprises the measurement data collection from the $\Delta\Sigma$ modulators *Analog Devices AD7401A*. The digital filters (decimators) were implemented as Sinc3 filters and were implemented on the FPGA. The FPGA also incorporates the low-level control of the power converters, whereas the high level control was implemented in the processor.

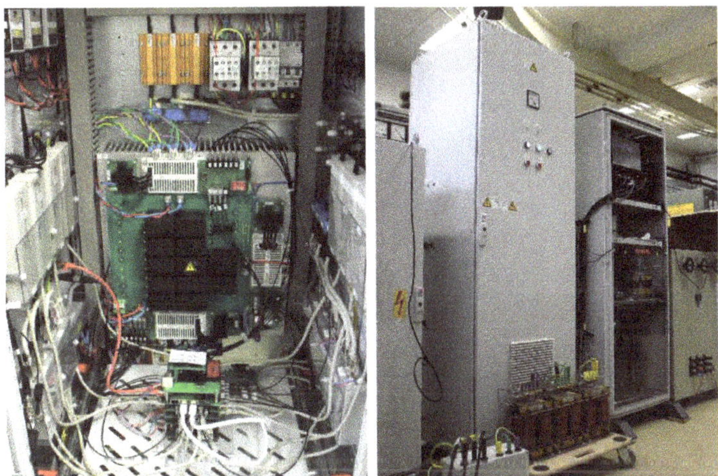

Figure 7. Photos of the experimental system.

The voltage differential probe measuring the dc-link voltage (*Tektronix P5200*) and the line-to-line voltage (*Testec TT-SI-9001*) were connected directly to the terminals of the IGBT module. The ac-side current was measured in phase 1 using a current probe (*Agilent N2782B*). The dc-side current and the rectifier current were measured with Rogowski coils (*PEM CWT03LF*), thus the dc component was excluded in the presented results. The gate signals of the top and bottom IGBTs of phase 1 were measured at the input terminals of the gate driver (*Testec TT-SI-9001*).

The data acquisition was performed using an eight channel oscilloscope *Teledyne Lecroy HDO8108*, which is depicted on the right side of Figure 7. By using a high sampling rate of 250 MS/s, 20 MHz analog input filters for the channels, and a 12 bit quantization, a good time and amplitude resolution of the measured waveforms was provided. The measured waveforms were postprocessed in *Matlab* to calculate the spectra.

4.2. Results for AD-PWM Model

To validate the correct derivation of (26), the equation was numerically evaluated for an example case with a duty cycle spectrum that contains frequency components at the fundamental frequency and at the fifth harmonic. The equivalent time-domain waveform is

$$x(t) = 0.5 \cdot \cos(2\pi f_0 t) + 0.5 \cdot \cos(2\pi \cdot 5 f_0 t), \tag{34}$$

with a fundamental frequency of $f_0 = 50\,\text{Hz}$. The sampling frequency of the ZOH is $f_s = 4\,\text{kHz}$ and the carrier frequency is $f_{sw} = 2\,\text{kHz}$. The first graph of Figure 8 shows the amplitude spectrum resulting from the application of (26) for the first 100 harmonics of f_0. The indexes of summation in the evaluation of (26) were limited to $-m_{max} \leq m \leq m_{max}$ and $1 \leq n \leq n_{max}$, with $m_{max} = 3$ and $n_{max} = 15$.

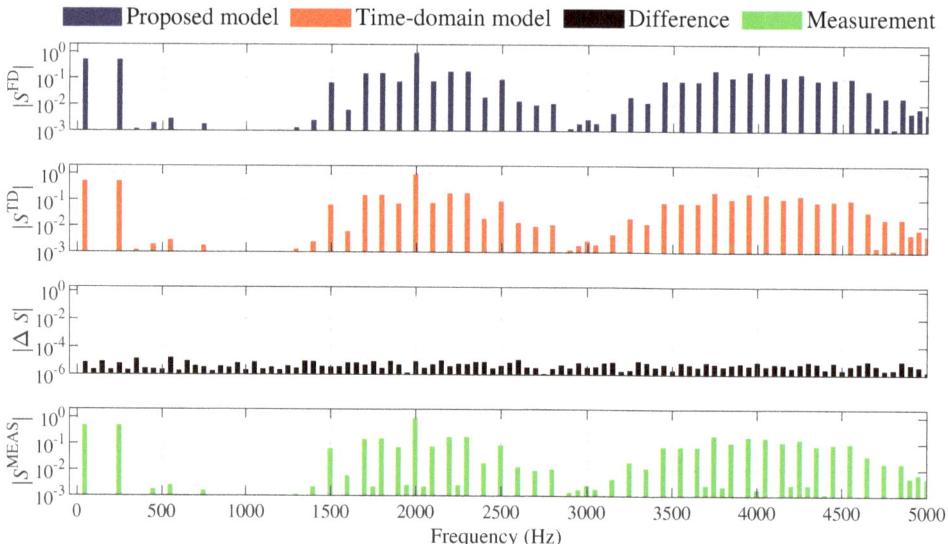

Figure 8. Amplitude spectra of the proposed frequency-domain model (first graph), the time-domain simulation (second graph), their difference based on the complex spectra (third graph), and the amplitude spectra from measurement results (fourth graph).

The second graph shows the spectrum from a time-domain simulation in *Simulink*, resulting from a Fast Fourier transform (FFT) of the results calculated in *Matlab*. The third graph depicts the difference of spectra with black bars, with

$$|\Delta S(f)| = |S^{\mathrm{TD}}(f) - S^{\mathrm{FD}}(f)|, \tag{35}$$

where TD and FD identify the time-domain simulation and the frequency-domain calculation, respectively. The differences in the spectra are much lower than 10^{-3}. Because the spectral difference is shown for the complex spectra, the validation also ensures that the phase information is correct. The fourth graph of Figure 8 presents the amplitude spectra $|S^{\mathrm{MEAS}}(f)|$ resulting from measurement of the gate-driver signals, further validating the proposed model.

4.3. Results for AD-PWM Model with DTFT Input

A numerical comparison of the two models presented in Figure 6a,b was performed, using the model by Song and Sarwate ([11], (61)) for ND-PWM and the proposed model for AD-PWM in (26). The input is the spectrum of the sampled duty cycle, as shown in Figure 9. It comprises a fundamental frequency component with an amplitude of $|D_{,1}| = 0.7$ and a seventh harmonic component with an amplitude of $|D_{,7}| = 0.1$. The carrier to fundamental ratio is $f_{\mathrm{sw}}/f_{d0} = 3000/50 = 60$.

The amplitude spectrum of the switching function is shown in the graphs on the left side of Figure 10, while varying the parameters $m_{\max}$ and $n_{\max}$. The first graph compares the frequency-domain model with a time-domain model for $m_{\max} = 1$, $n_{\max} = 7$, $k_{\max} = 150$. Large deviations between the complex spectra are visible in the second graph. Increasing $m_{\max}$ and $n_{\max}$ leads only to a minor reduction of the deviations, because of the low number of harmonics $k_{\max}$ considered in the duty-cycle spectrum (third and fourth graph). With these parameters, the second and third sideband groups depicted in Figure 9 are neglected, resulting in a faulty calculation of the first carrier group for the switching function spectrum.

The method demonstrated in Figure 6b converts the DTFT of the duty-cycle first into a CTFT, removing the high-frequency components shown in Figure 9 before application to the AD-PWM model. The right side of Figure 10 presents the evaluation of this model for the same parameters as for the ND-PWM model. Even though the modeling approach seems to be more complex, a high accuracy is reached for relatively low numbers of $k_{\max}$. Due to the rectangular filter, the duty-cycle components requiring consideration in the PWM model are limited to $f_c/2$. This numerical example demonstrates that model b) requires a lower number of harmonics $k_{\max}$ in the duty-cycle to reach the same accuracy as the AD-PWM model, which is crucial for the numerical solution process of the overall system.

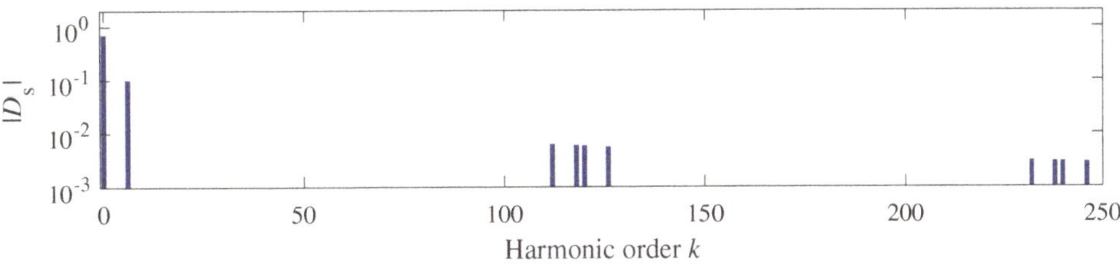

Figure 9. Spectrum of the sampled and held duty cycle with two components at $k = 1$ and $k = 7$. Due to sampling, components around multiples of the $f_c/f_0 = 60$ are present [27].

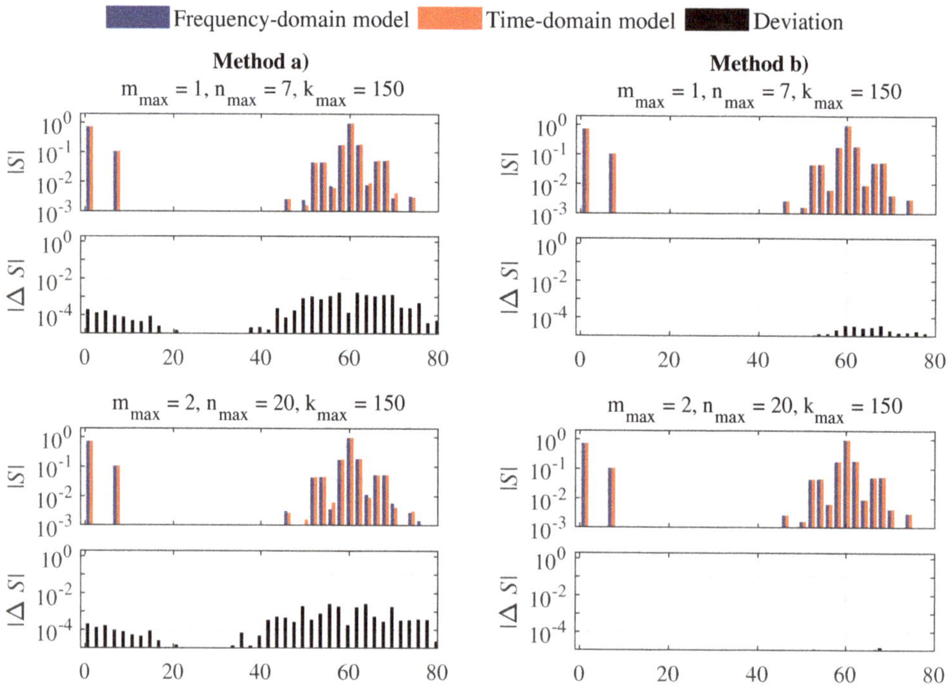

Figure 10. Evaluation of the models depicted in Figure 6a (left graphs) and Figure 6b (right graphs). The parameters m_{max} and n_{max} are varied from top to bottom [27].

4.4. Results for the VSI Model

The evaluation of the VSI model was conducted for an example case with the parameters listed in Table 1. The PI controller was designed using the technical optimum. Harmonic components were introduced by the nonreactive dc current source with frequencies of $n \cdot 6 \cdot f_g, n \in \mathbb{N}$ as an example of a three-phase diode rectifier fed from a symmetrical 400 V grid.

The frequency-domain model was evaluated with $k_{max} = 70$, $m_{max} = 1$, and $n_{max} = 7$. The magnitude spectra are depicted in Figure 11, comparing the results from the frequency-domain model to results from a time-domain simulation. As shown in Figure 1, the spectrum I_{rec} and the dc-component of the dc-link voltage in the frequency-domain model are input signals of the frequency-domain model and were selected to be the same values as in the time-domain simulation. Low-frequency components were introduced into the duty cycle spectrum through the control loop of the ac-side currents and the feedback of the dc-link voltage. These components are a product of the disturbance current of the diode rectifier, which are propagated to the sensed dc-link voltage and the sensed ac-side currents. Because the PWM process is nonlinear, the harmonic components of the duty cycle spectrum create additional base-band and side-band harmonics in the switching function spectrum.

The results of the two models have a high rate of conformity, which is demonstrated by the low deviations of the complex spectra shown in Figure 12. Minor deviations visible around the switching frequency can be explained by the low number of considered harmonics, as demonstrated by the previous examination shown in Figure 10. For instance, the components in the duty cycle spectrum around the first carrier-sideband group would cancel out with components of the second carrier-sideband group, indicating that an improvement of the method to enable a higher number of considered harmonics is desirable.

A comparison of the frequency-domain results with measurement results is presented in Figure 13, confirming the pattern of the calculated spectra. The VSI was fed from a three-phase diode rectifier and the spectrum I_{rec} and the dc-component of the dc-link voltage were used as input signals for the frequency-domain model. The deviations between the frequency-domain model and measurements are higher in comparison to the deviations from the simulation results, indicating the influence of neglected effects in the models such as converter dead time.

Table 1. Parameters of the example system.

Parameter	Symbol	Values
Fundamental frequency	f_0	50 Hz
Switching frequency	$f_{sw} = 0.5 \cdot f_s$	3 kHz
AC-side resistance	R_{ac}	5 Ω
AC-side inductance	L_{ac}	20 mH
DC-link capacitance	C_{dc}	480 µF
Grid voltage (line-to-line, rms) (rectifier)	U_g	400 V
Grid frequency (rectifier)	f_g	50 Hz
Grid-side inductance (rectifier)	L_g	260 µH
Proportional gain (current control)	K_P	20 V/A
Integrator time constant controller	T_I	4 ms

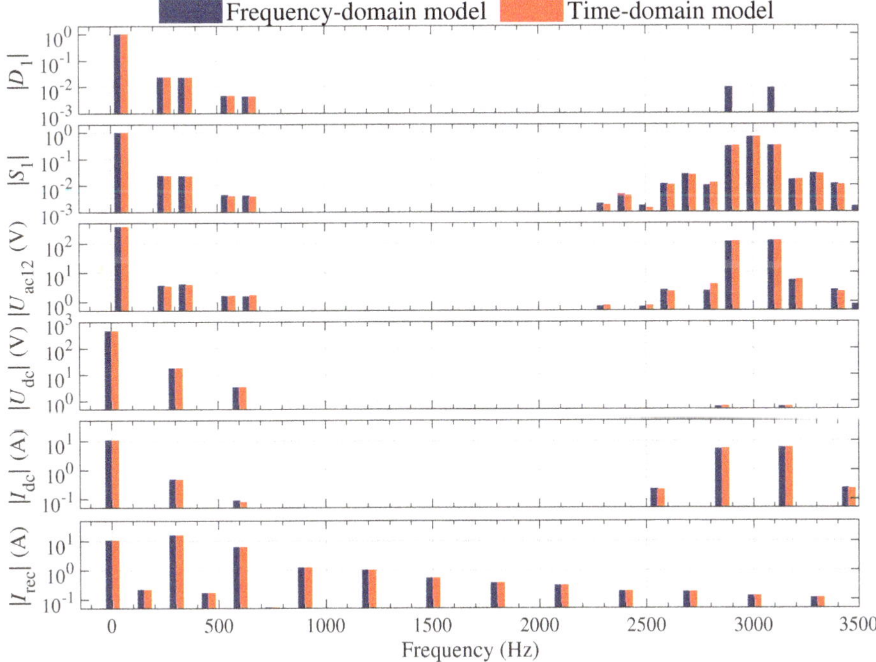

Figure 11. Comparison of the results for closed-loop control using the frequency-domain model (blue) and the time-domain model (red).

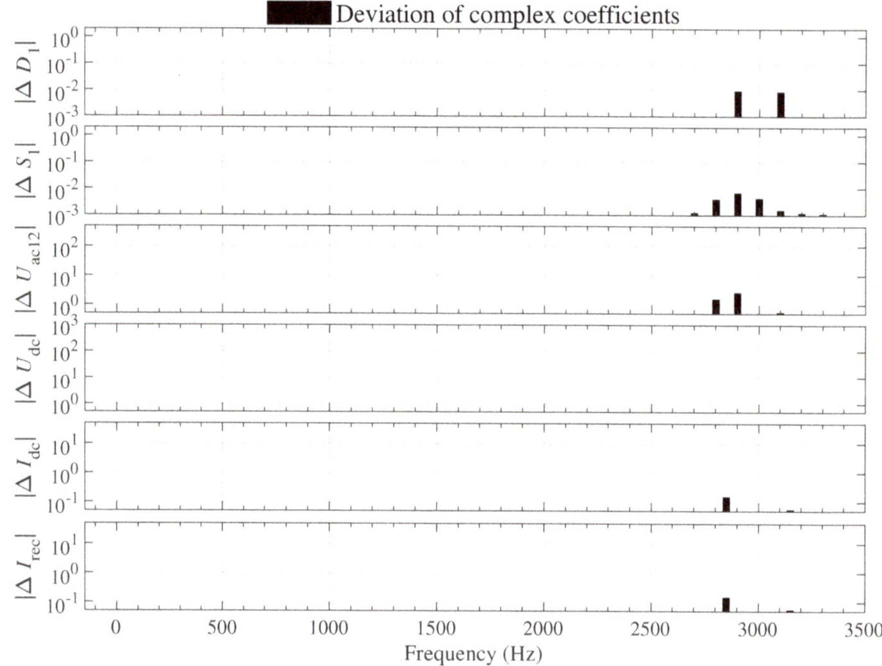

Figure 12. Deviation of the complex spectra of the frequency-domain model from the time-domain model presented in Figure 11 for closed-loop control.

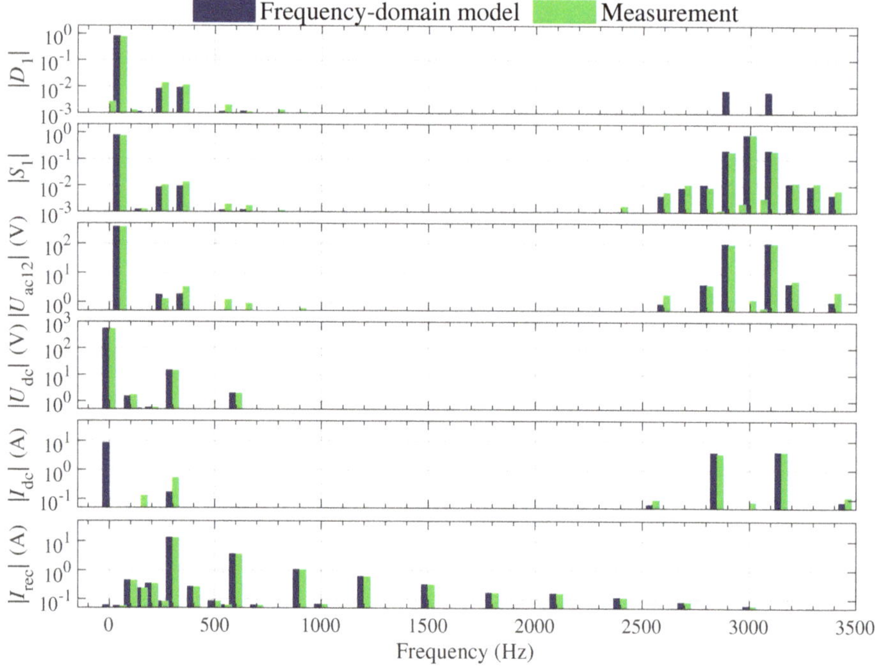

Figure 13. Comparison of frequency-domain model and measurements for a three-phase VSI with closed-loop control for Case I.

5. Conclusions

The control PWM signals and the electrical system of a VSI form a closed-loop system. A system model was developed in the frequency-domain that incorporates the interaction of the closed-loop control, the PWM process, and the power electronic devices as ideal switches. Although past methods in the literature demonstrate unidirectional descriptions for the individual parts of the converter system, the approach posed here includes their mutual dependencies and nonlinear interactions. The overall nonlinear equation system was numerically and simultaneously solved. The analysis showed how the propagation of harmonic components through the control loop alters the mutually coupled spectra of the duty cycle and the switching function.

A PWM model that considers multiple-frequency duty cycles is necessary to cover these interactions. This paper demonstrates the derivation of a revised analytic frequency-domain model for AD-PWM that is valid for multiple-frequency input signals. Low deviations between the numerical results from this model and those of time-domain simulations, as well as those of measurements, support the proposed equations. This paper illustrated the approximation of a regularly-sampled PWM model with a continuous-time input signal by a naturally-sampled PWM model that is extended with a sample-and-hold unit at its input. For the case of discrete-time input signals, an inverse sample-and-hold process in combination with a regularly-sampled PWM can be applied. It was demonstrated that a smaller number of considered harmonics is required for this approach to achieve the same accuracy as a naturally-sampled PWM model.

Application of the presented method is recommended when the interaction of the harmonic components cannot be neglected and when the base band harmonics and the switching band harmonics cannot be clearly separated. Due to the detailed system description, the computational effort is generally high and increases further with the number of considered harmonics and the number of system variables. An enhancement of the computational speed is desirable, when large numbers of considered harmonics and multiple nonlinear effects are present. For this, further development of the system description and an extensive study of solvers and an analytical Jacobian matrix are recommended.

Author Contributions: Conceptualization, methodology, validation, formal analysis, writing—original draft preparation, M.J.; funding acquisition, project administration, supervision, writing—review and editing, A.M.; All authors have read and agreed to the published version of the manuscript.

Funding: This research was funded in part by the Volkswagen Foundation's grant "Niedersächsisches Vorab 2014" within the project AMSES (Aggregated Models for the Simulation of Electromechanical Power Systems). The publication of this article was funded by the Open Access Fund of Leibniz Universität Hannover.

Conflicts of Interest: The authors declare no conflict of interest.

Abbreviations

The following abbreviations are used in this manuscript:

ADC	Analog-to-digital converter
AD-PWM	Asymmetrical regularly-sampled double-edge PWM
CTFT	Continuous-time Fourier transform
DTFT	Discrete-time Fourier transform
FFT	Fast Fourier transform
LE	Leading edge
LE-PWM	Leading-edge PWM
ND-PWM	Naturally-sampled double-edge PWM
PWM	Pulse-width modulation
TE	Trailing edge
TE-PWM	Trailing-edge PWM
VSI	Voltage-source inverter
ZOH	Zero-order hold

Appendix A

This appendix provides the derivation of a frequency-domain expression for AD-PWM in (26) as a revision of the model presented in [11]. When using double-edge PWM, the placement of both edges within the switching period depends on the duty cycle. Sampling the duty cycle at the beginning and the center of the carrier period results in AD-PWM. Its analysis can be simplified by regarding the rising edge as a trailing-edge PWM (TE-PWM) signal and the falling edge as a leading-edge PWM (LE-PWM) signal [31]. This approach is illustrated in Figure A1, where the AD-PWM process is depicted for one switching period and a positive duty cycle. Subfigure a shows the comparison of a sampled duty cycle (red dashed line) with a triangular carrier signal (black solid line). The duty cycle is sampled and held at the beginning of the switching period for the leading edge (LE) and at the center of the period for the trailing edge (TE). The resulting output can be composed of:

(1) A rectangular signal s_c with constant duty cycle of 50 % (Subfigure d): In accordance with Song and Sarwate [11], the signal $s_c(t)$ is defined as a rectangular wave

$$s_c(t) = \begin{cases} 1 & \text{for} \quad mT_{sw} \le t < (m+1/2)T_{sw} \\ -1 & \text{for} \quad (m+1/2)T_{sw} \le t < (m+1)T_{sw} \end{cases}, \tag{A1}$$

where $m \in \mathbb{Z}$, $T_{sw} = 1/f_{sw}$, and t is the time. The signal in (Subfigure d) is right-shifted by $T_{sw}/4$.

(2) A rectangular signal s_{dLE} that is determined by the LE (Subfigure e): Its rising edge results from a comparison of the duty-cycle sampled at the beginning of the switching period $d[mT_{sw}]/2$ with a LE carrier signal (Subfigure b). The original duty cycle is plotted in light red and the halved value is depicted in red. The carrier signal is left-shifted by $T_{sw}/4$ (original carrier depicted in gray and shifted carrier in black). The falling edge of s_{dLE} is fixed to $T_{sw}/4 + mT_{sw}$.

(3) A rectangular signal s_{dTE} that is determined by the TE (Subfigure f): Its falling edge results from a comparison of the duty-cycle sampled at the center of the switching period $d[(m+1/2)T_{sw}]/2$ with a TE carrier signal that is right-shifted by $T_{sw}/4$ (Subfigure c). The rising edge of s_{dTE} is fixed to $3T_{sw}/4 + mT_{sw}$.

Their superposition yields the switching function for AD-PWM, with

$$\begin{aligned} s_{AD}(t,d) = s_c(t - T_{sw}/4) + s_{dLE}(t + T_{sw}/4;\ d[mT_{sw}]/2)... \\ + s_{dTE}(t - T_{sw}/4;\ d[(m+1/2)T_{sw}]/2). \end{aligned} \tag{A2}$$

The sampling points of the duty cycle $[mT_{sw}]$ and $[(m+1/2)T_{sw}]$ are included to stress the difference between the TE and the LE. This equation differs from the result of Song and Sarwate ([11], (56)), where the different sampling instances for LE and TE are not represented. The Fourier transform of the switching function in (A2) results in

$$\begin{aligned} S_{AD}(f,d) = e^{-j\pi f T_{sw}/2} \cdot S_c(f) + e^{-j\pi f T_{sw}/2} \cdot S_{dTE}(f, d[(m+1/2)T_{sw}]/2)... \\ + e^{+j\pi f T_{sw}/2} \cdot S_{dLE}(f, d[mT_{sw}]/2). \end{aligned} \tag{A3}$$

In the following equations, explicit frequency-domain expressions of the three components in (A3) are derived. The spectrum of the rectangular function of constant duty cycle is given in (24).

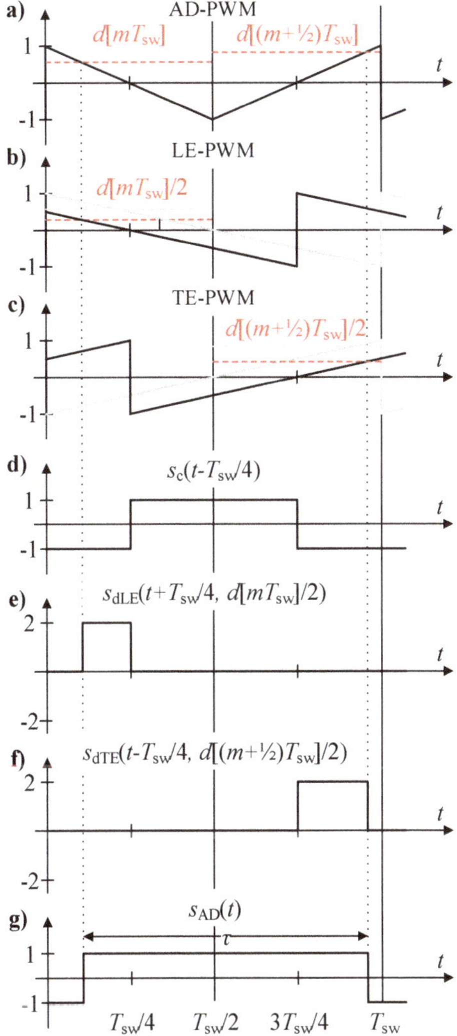

Figure A1. Derivation of switching function for AD-PWM: (**a**) Comparison of an AD-PWM carrier signal (black line) with a sampled duty cycle sampled at mT_{sw} and $(m+1/2)T_{\text{sw}}$ (red dashed line). (**b**) Equivalent description for the leading edge using a LE-PWM carrier signal left-shifted by $T_{\text{sw}}/4$. The duty cycle's value is halved and sampled at mT_{sw}. (**c**) Equivalent description for the trailing edge using a TE-PWM carrier signal right-shifted by $T_{\text{sw}}/4$. The duty cycle's value is halved and sampled at $(m+1/2)T_{\text{sw}}$. (**d**) Rectangular function with a duty cycle of 50 % right-shifted by $T_{\text{sw}}/4$. (**e**) Duty-cycle-dependent part for LE resulting from (**b**). (**f**) Duty-cycle-dependent part for TE resulting from (**c**). (**g**) Resulting switching function comprising (**d**) + (**e**) + (**f**) [27].

The two duty-cycle-dependent components can be expressed by using a Lagrange extension of the rectangular functions, where the TE component can be expressed as

$$
\begin{aligned}
S_{\text{dTE}}(f,d) = & \frac{1}{\mathrm{j}\pi f} \cdot \mathrm{e}^{-\mathrm{j}\pi f T_{\text{sw}}} \cdot \sum_{n=1}^{\infty} -\frac{(-\mathrm{j}\pi f T_{\text{sw}})^n}{n!} \cdot \sum_{m=-\infty}^{\infty} (d[mT_{\text{sw}}])^n \cdot \mathrm{e}^{-\mathrm{j}2\pi f m T_{\text{sw}}}, \\
& k \in \mathbb{Z}, n \in \mathbb{N} \quad .
\end{aligned}
\tag{A4}
$$

Similarly, the LE component results in

$$S_{\mathrm{dLE}}(f,d) = -S_{\mathrm{dTE}}(f,-d) \tag{A5}$$

$$= -\frac{1}{\mathrm{j}\pi f}\cdot \mathrm{e}^{-\mathrm{j}\pi f T_{\mathrm{sw}}}\cdot \sum_{n=1}^{\infty} -\frac{(\mathrm{j}\pi f T_{\mathrm{sw}})^n}{n!}\cdot \sum_{m=-\infty}^{\infty} (d[mT_{\mathrm{sw}}])^n \cdot \mathrm{e}^{-\mathrm{j}2\pi f m T_{\mathrm{sw}}}. \tag{A6}$$

In order to express the switching function spectrum as a function of the duty cycle spectrum $D(f) \longmapsto S_{\mathrm{AD}}(f)$, the time-domain expression $(d[mT_{\mathrm{sw}}])^n$ needs to be replaced by a frequency-domain expression. By acknowledging that exponentiation of a signal in the time domain relates to a repeated convolution in the frequency domain, the Fourier transform of the continuous-time duty-cycle is written as $D^{*n}(f)$, see (25). Because the duty-cycle for regularly-sampled PWM is a discrete-time signal, the discrete-time Fourier transform is applied, which results in

$$\sum_{m=-\infty}^{\infty} (d[mT_{\mathrm{sw}}])^n \cdot \mathrm{e}^{-\mathrm{j}2\pi f m T_{\mathrm{sw}}} = f_{\mathrm{sw}} \cdot \sum_{m=-\infty}^{\infty} D^{*n}(f - m f_{\mathrm{sw}}). \tag{A7}$$

The influence of the duty cycle's time shift of $T_{\mathrm{sw}}/2$ on the Fourier transform in (A7) yields

$$\sum_{m=-\infty}^{\infty} (d[(m+1/2)T_{\mathrm{sw}}])^n \cdot \mathrm{e}^{-\mathrm{j}2\pi f m T_{\mathrm{sw}}} = f_{\mathrm{sw}} \cdot \sum_{m=-\infty}^{\infty} \mathrm{e}^{\mathrm{j}\pi\cdot(f - m f_{\mathrm{sw}})T} \cdot D^{*n}(f - m f_{\mathrm{sw}}). \tag{A8}$$

Its application to (A4) allows for the description of the TE component for AD-PWM, with

$$\begin{aligned} &S_{\mathrm{dTE}}(f, d[(m+1/2)T_{\mathrm{sw}}]/2) \\ &= \frac{1}{\mathrm{j}\pi f}\cdot \mathrm{e}^{-\mathrm{j}\pi f T_{\mathrm{sw}}}\cdot \sum_{n=1}^{\infty} -\frac{(-\mathrm{j}\pi f T_{\mathrm{sw}})^n}{n!}\cdot \sum_{m=-\infty}^{\infty} (d[(m+1/2)T_{\mathrm{sw}}]/2)^n \cdot \mathrm{e}^{-\mathrm{j}2\pi f m T_{\mathrm{sw}}} \\ &= \mathrm{e}^{-\mathrm{j}\pi f T_{\mathrm{sw}}}\cdot \sum_{m=-\infty}^{\infty}\sum_{n=1}^{\infty} \frac{(-\mathrm{j}\pi f T_{\mathrm{sw}})^{n-1}}{2^n \cdot n!}\cdot D^{*n}(f - m f_{\mathrm{sw}}) \cdot \mathrm{e}^{\mathrm{j}\pi\cdot(f - m f_{\mathrm{sw}})T_{\mathrm{sw}}}. \end{aligned} \tag{A9}$$

Similarly, the LE component is expressed as

$$\begin{aligned} &S_{\mathrm{dLE}}(f, d[mT_{\mathrm{sw}}]/2) \\ &= -\frac{1}{\mathrm{j}\pi f}\cdot \mathrm{e}^{-\mathrm{j}\pi f T_{\mathrm{sw}}}\cdot \sum_{n=1}^{\infty} -\frac{(\mathrm{j}\pi f T_{\mathrm{sw}})^n}{n!}\cdot \sum_{m=-\infty}^{\infty} (d[kT_{\mathrm{sw}}]/2)^n \cdot \mathrm{e}^{-\mathrm{j}2\pi f m T_{\mathrm{sw}}} \\ &= \mathrm{e}^{-\mathrm{j}\pi f T_{\mathrm{sw}}}\cdot \sum_{m=-\infty}^{\infty}\sum_{n=1}^{\infty} \frac{(\mathrm{j}\pi f T_{\mathrm{sw}})^{n-1}}{2^n \cdot n!}\cdot D^{*n}(f - m f_{\mathrm{sw}}). \end{aligned} \tag{A10}$$

Applying the results from (A9) and (A10) to (A3) results in

$$\begin{aligned} &S_{\mathrm{AD}}(f, D) = \mathrm{e}^{-\mathrm{j}\pi f T_{\mathrm{sw}}/2}\cdot S_{\mathrm{c}}(f) \ldots \\ &+ \mathrm{e}^{-\mathrm{j}\pi f T_{\mathrm{sw}}/2}\cdot \mathrm{e}^{-\mathrm{j}\pi f T_{\mathrm{sw}}}\cdot \sum_{m=-\infty}^{\infty}\sum_{n=1}^{\infty} \frac{(-\mathrm{j}\pi f T_{\mathrm{sw}})^{n-1}}{2^n \cdot n!}\cdot D^{*n}(f - m f_{\mathrm{sw}}) \cdot \mathrm{e}^{\mathrm{j}\pi\cdot(f - m f_{\mathrm{sw}})T} \ldots \\ &+ \mathrm{e}^{+\mathrm{j}\pi f T_{\mathrm{sw}}/2}\cdot \mathrm{e}^{-\mathrm{j}\pi f T_{\mathrm{sw}}}\cdot \sum_{m=-\infty}^{\infty}\sum_{n=1}^{\infty} \frac{(\mathrm{j}\pi f T_{\mathrm{sw}})^{n-1}}{2^n \cdot n!}\cdot D^{*n}(f - m f_{\mathrm{sw}}), \end{aligned} \tag{A11}$$

which can be summarized in the final form of (26).

References

1. Bierhoff, M.; Fuchs, F. DC-Link Harmonics of Three-Phase Voltage-Source Converters Influenced by the Pulsewidth-Modulation Strategy—An Analysis. *IEEE Trans. Ind. Electron.* **2008**, *55*, 2085–2092. [CrossRef]

2. McGrath, B.; Holmes, D. A General Analytical Method for Calculating Inverter DC-Link Current Harmonics. *IEEE Trans. Ind. Appl.* **2009**, *45*, 1851–1859. [CrossRef]
3. Meyer, R.; Zlotnik, A.; Mertens, A. Fault Ride-Through Control of Medium-Voltage Converters With LCL Filter in Distributed Generation Systems. *IEEE Trans. Ind. Appl.* **2014**, *50*, 3448–3456. [CrossRef]
4. Fuchs, F. Converter Control for Wind Turbines when Operating inWeak Grids Containing Resonances. Ph.D. Thesis, Leibniz Universität Hannover, Hanover, Germany, 2017.
5. Odavic, M.; Sumner, M.; Zanchetta, P.; Clare, J.C. A Theoretical Analysis of the Harmonic Content of PWM Waveforms for Multiple-Frequency Modulators. *IEEE Trans. Power Electron.* **2010**, *25*, 131–141. [CrossRef]
6. Rahmani, S.; Al-Haddad, K.; Kanaan, H.Y. Two PWM techniques for single-phase shunt active power filters employing a direct current control strategy. *IET Power Electron.* **2008**, *1*, 376–385. [CrossRef]
7. Bowes, S.; Bird, B. Novel approach to the analysis and synthesis of modulation processes in power convertors. *Proc. Inst. Electr. Eng.* **1975**, *122*, 507–513. [CrossRef]
8. Holmes, G. A Generalised Approach to the Modulation and Control of Hard Switched Converters. Ph.D. Thesis, Monash University, Melbourne, Australia, 1997.
9. Holmes, G.; Lipo, T.A. *Pulse Width Modulation for Power Converters: Principles and Practice*; Wiley-IEEE Press: Hoboken, NJ, USA, 2003.
10. Cox, S.M. Voltage and current spectra for a single-phase voltage source inverter. *IMA J. Appl. Math.* **2009**, *74*, 782–805. [CrossRef]
11. Song, Z.; Sarwate, D.V. The frequency spectrum of pulse width modulated signals. *Signal Process.* **2003**, *83*, 2227–2258. [CrossRef]
12. Kostic, D.J.; Avramovic, Z.Z.; Ciric, N.T. A New Approach to Theoretical Analysis of Harmonic Content of PWM Waveforms of Single- and Multiple-Frequency Modulators. *IEEE Trans. Power Electron.* **2013**, *28*, 4557–4567. [CrossRef]
13. Mouton, H.D.T.; McGrath, B.; Holmes, D.G.; Wilkinson, R.H. One-Dimensional Spectral Analysis of Complex PWM Waveforms Using Superposition. *IEEE Trans. Power Electron.* **2014**, *29*, 6762–6778. [CrossRef]
14. Thakur, S.S.; Odavic, M.; Allu, A.; Zhu, Z.Q.; Atallah, K. Theoretical Harmonic Spectra of PWM Waveforms Including DC Bus Voltage Ripple—Application to a Low-Capacitance Modular Multilevel Converter. *IEEE Trans. Power Electron.* **2020**, *35*, 9291–9305. [CrossRef]
15. Milovanović, S.; Dujić, D. Comprehensive Spectral Analysis of PWM Waveforms With Compensated DC-Link Oscillations. *IEEE Trans. Power Electron.* **2020**, *35*, 12898–12908. [CrossRef]
16. Wester, G.W.; Middlebrook, R.D. Low-Frequency Characterization of Switched dc-dc Converters. *IEEE Trans. Aerosp. Electron. Syst.* **1973**, *AES-9*, 376–385. [CrossRef]
17. Cuk, S. Modelling, Analysis, and Design of Switching Converters. Ph.D. Thesis, California Institute of Technology, Pasadena, CA, USA, 1977.
18. Middlebrook, R.D.; Ćuk, S. A general unified approach to modelling switching-converter power stages. *Int. J. Electron.* **1977**, *42*, 521–550. [CrossRef]
19. Erickson, R.W.; Maksimovic, D. *Fundamentals of Power Electronics*, 2nd ed.; Springer US: New York, NY, USA, 2001.
20. Maksimovic, D.; Stankovic, A.M.; Thottuvelil, V.J.; Verghese, G.C. Modeling and simulation of power electronic converters. *Proc. IEEE* **2001**, *89*, 898–912. [CrossRef]
21. Maksimovic, D.; Zane, R. Small-Signal Discrete-Time Modeling of Digitally Controlled PWM Converters. *IEEE Trans. Power Electron.* **2007**, *22*, 2552–2556. [CrossRef]
22. Hiti, S. Modeling and Control of Three-Phase PWM Converters. Ph.D. Thesis, Virginia Tech, Blacksburg, VA, USA, 1995.
23. Sun, J. Small-Signal Methods for AC Distributed Power Systems—A Review. *IEEE Trans. Power Electron.* **2009**, *24*, 2545–2554. [CrossRef]
24. Corradini, L.; Maksimovic, D.; Mattavelli, P.; Zane, R. *Digital Control of High-Frequency Switched-Mode Power Converters*; Wiley-IEEE Press: Hoboken, NJ, USA, 2015.
25. Almér, S.; Jönsson, U. Harmonic analysis of pulse-width modulated systems. *Automatica* **2009**, *45*, 851–862. [CrossRef]
26. Yue, X.; Boroyevich, D.; Lee, F.C.; Chen, F.; Burgos, R.; Zhuo, F. Beat Frequency Oscillation Analysis for Power Electronic Converters in DC Nanogrid Based on Crossed Frequency Output Impedance Matrix Model. *IEEE Trans. Power Electron.* **2018**, *33*, 3052–3064. [CrossRef]

27. John, M. Frequency-Domain Modeling of Harmonic Interactions in Pulse-Width Modulated Voltage-Source Inverter Drives. Ph.D. Thesis, Leibniz Universität Hannover, Hanover, Germany, 2019.
28. Candy, J.C.; Tmes, G.C. *Oversampling Delta-Sigma Data Converters: Theory, Design, and Simulation*; IEEE Press, Wiley-Interscience: Piscataway, NJ, USA, 1992.
29. Buso, S.; Mattavelli, P. *Digital Control in Power Electronics*, 2nd ed.; Morgan & Claypool: San Rafael, CA, USA, 2015.
30. Arrillaga, J.; Smith, B.C.; Watson, N.R.; Wood, A.R. *Power System Harmonic Analysis*; Wiley & Sons Ltd.: Chichester, UK, 1997.
31. Black, H.S. *Modulation Theory*; Van Nostrand: Princeton, NJ, USA, 1953.

Article

Coordinated Control of Active and Reactive Power Compensation for Voltage Regulation with Enhanced Disturbance Rejection Using Repetitive Vector-Control

Felipe J. Zimann [1,*], Eduardo V. Stangler [2], Francisco A. S. Neves [2], Alessandro L. Batschauer [1] and Marcello Mezaroba [1]

1 Department of Electrical Engineering, Santa Catarina State University, Joinville 89219-710, Brazil; alessandro.batschauer@udesc.br (A.L.B.); marcello.mezaroba@udesc.br (M.M.)
2 Department of Electrical Engineering, Universidade Federal de Pernambuco, Recife 50740-530, Brazil; eduardo.vasconcelosstangler@ufpe.br (E.V.S.); francisco.neves@ufpe.br (F.A.S.N.)
* Correspondence: felipe.zimann@edu.udesc.br

Received: 9 April 2020; Accepted: 13 May 2020; Published: 2 June 2020

Abstract: Voltage profile is one of many aspects that affect power quality in low-voltage distribution feeders. Weak grids have a typically high line impedance which results in remarkable voltage drops. Distribution grids generally have a high R/X ratio, which makes voltage regulation with reactive power compensation less effective than in high-voltage grids. Moreover, these networks are more susceptible to unbalance and harmonic voltage disturbances. This paper proposes an enhanced coordinated control of active and reactive power injected in a distribution grid for voltage regulation. Voltage drop mitigation was evaluated with power injection based on local features, such loads and disturbances of each connection. In order to ensure disturbances rejection like harmonic components in the grid voltages, a repetitive vector-control scheme was used. The injection of coordinated active and reactive power with the proposed control algorithm was verified through simulations and experiments, demonstrating that it is a promising alternative for voltage regulation in weak and low-voltage networks subject to inherent harmonic distortion.

Keywords: effective voltage regulation; generalized delayed signal cancellation; harmonic distortion; power quality; repetitive controller

1. Introduction

Low-voltage (LV) distribution systems may suffer from the reduction of voltage amplitude, especially at points far away from the supply locations [1]. A significant voltage drop occurs when both the feeder and line impedance have high values and the drop voltages on these elements assume a considerable value in relation to the voltage delivered to the consumers. In such cases, the grid is called a weak grid [2]. Any branch of a low voltage weak grid may suffer from this effect. Other factors may also influence the quality of the power supplied, such as the voltage distortion at the point of common connection (PCC) produced by non-linear loads that drain currents with a high harmonic content [3]. In weak grids, harmonic distortions become more apparent compared to the fundamental component. This occurs primarily because line impedance is directly proportional to the frequency and the nonlinear current flowing through the branch produces non-fundamental frequency voltage drops.

To address the voltage drop issue, reactive power injection is used to regulate the voltage at specific points located on the grid with traditional reactive control and distribution static synchronous compensators (DSTATCOM). However, this method presents one major drawback: reactive power control is not as effective as in high-voltage grids in consequence of the R/X ratio being usually high in LV grids [4]. On the other hand, several related approaches were introduced in [5–8] which confirm the effectiveness of active-based voltage regulation. Battery energy storage systems are the most commonly used equipment for storing energy [9–12]. However, deep discharges, high drain current and high operating temperatures are some issues that reduce the lifespan of those systems [13].

In recent works [4,8,14–16], the reactive power control was redesigned to work with active power. Generally, DSTATCOMs use only fundamental–frequency current injection for rms voltage regulation [16,17]. However, the grid voltages and their harmonic distortions are disturbances to the current control. Even if the DSTATCOM reference currents are perfectly sinusoidal and balanced, the grid voltages may affect the controllers' performance, leading to steady–state errors, distorted injected currents and even instability if not properly addressed during the control design. Furthermore, additional functions can be included in the DSTATCOM control, like compensating load unbalance and harmonic currents. In this case, the controller reference current must have high gain for fundamental frequency positive- and negative-sequence components and for all harmonic components that are expected in the load currents. In order to comply with these requirements, PI conventional controllers with very high bandwidth would be necessary and multiple-frequency controllers based on proportional-resonant actions have been thoroughly investigated, where the latter solution have shown to be preferable. In this sense, repetitive controllers also appear as good candidates as presented in [18–25].

In general, the traditional repetitive control (RC) techniques have a slow dynamic response and a transient performance that is not suitable for controlling the instantaneous variables of pulse-width modulated (PWM) converters. As a result, these RCs are used as plug-in controllers and work in addition to classical controllers, usually proportional-integral (PI) or proportional-resonant (PR) type [18–20]. In other cases, repetitive controllers are used for steady state error elimination, where a fast dynamic performance is not necessary. However, complex repetitive vector-control schemes recently proposed have demonstrated better stability and performance characteristics, in comparison with scalar repetitive controllers [26,27]. The use of complex structures of repetitive controllers may be suitable for applications where the controller performance is required in a family of harmonic components [27].

In [21] a conventional RC is used as a plug-in controller capable of regulating all integer harmonic components of a reference signal. This means that, if the reference signal waveform is composed of the fundamental-frequency (f_o) component together with any set of integer harmonic components, then the controller ensures zero steady-state error. Furthermore, if the system is subjected to integer harmonic disturbances, the controller rejects them in steady-state. For implementing this controller, all N samples of the signal error in the last fundamental period are necessary ($N = f_s/f_o$, where f_s is the sampling frequency). In [22,23] a RC designed for regulating odd harmonic components is used since even harmonic components are rarely found in electrical systems. Such selectivity allows the controller to use fewer samples of the signal error and also ensure a faster dynamic response, at the expense of not a ensuring zero steady-state error for even harmonic components. The mentioned idea was extended to allow the design of RCs for controlling a specific group of harmonic components, like $6k \pm 1$, $k \in \mathbb{N}$ [24], or $nk \pm m$, where n and m are fixed natural parameters and $k = 1, 2, 3, \ldots$ [25]. For bigger values of parameter n, less harmonic components are regulated, but faster dynamic responses are expected. The RCs described so far can be classified as scalar repetitive controllers. When applied to three-phase systems, one controller should be used to regulate each phase component or, alternatively, after a Clarke coordinate transformation, two of these control structures can be used in the α and β components [28].

In this paper, we propose a voltage regulation scheme with enhanced disturbance rejection for LV grids using repetitive vector control. The coordinated control of active and reactive power ensures effectiveness in weak grids, while space-vector repetitive controller (SV-RC) design rejects disturbances caused by current distortion and harmonic voltages on point of common coupling.

A space-vector controller has a unique structure that facilitates the control design, allows regulating a positive-sequence harmonic component without having to control the negative-sequence component of the same order and improves dynamic performance. Furthermore, our proposed active and reactive power coordinated control determines current references for power injection based on grid structure regarding the R/X ratio. As a result, the lifespan and autonomy of energy storage systems can be improved. It is worth mentioning that such N control technique can be adapted and then employed to any energy storage system, especially in remote electric-vehicle (EV) charging stations, in order to provide voltage regulation as an ancillary service [29]. Thus, depending on the state-of-charge of the batteries connected, it is possible to provide an effective voltage regulation capability with a minimum utilization of the stored energy of the EVs or any other storage system.

The paper is organized as follows: Section 2 presents the control and disturbance model, the SV-RC design focused on fundamental component control and disturbance rejection. Section 3 describes the active and reactive coordinate control as well as the effective voltage and power reading method. Section 4 presents the experimental results demonstrated by a prototype built in the laboratory. Finally, conclusions are presented in Section 5.

2. General Approach

The voltage regulation scheme determines, from the predefined voltage values, the currents to be injected by DSTATCOM. These references are determined based on the power management algorithm described in this section. The repetitive vector-control is used to ensure an effective control of currents, rejecting disturbances caused by imbalances and harmonic voltages on PCC. The proposed voltage regulation system with SV-RC is shown in Figure 1 where linear and non-linear loads, power grid and inverter are all connected to the PCC. An independent power source, a generic energy storage system (ESS), provides dc-link voltage regulation and the required power for the converter operation.

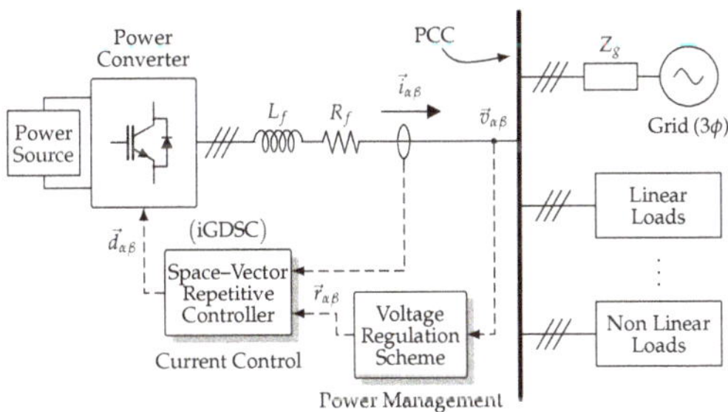

Figure 1. Voltage regulation system with SV-RC controller.

The converter output current $\vec{i}_{\alpha\beta}$ has two input variables: converter duty cycle $\vec{d}_{\alpha\beta}$ and the voltage at the PCC $\vec{v}_{\alpha\beta}$, which is considered a disturbance to the control model, as shown in Figure 2. The model equation can be obtained through Clarke coordinate transformation $abc - \alpha\beta$, as shown in (1):

$$\vec{i}_{\alpha\beta} = \underbrace{\left(\frac{E}{sL_f + R_f}\right)}_{G_{id}(s)} \cdot \vec{d}_{\alpha\beta} - \underbrace{\left(\frac{1}{sL_f + R_f}\right)}_{G_d(s)} \cdot \vec{v}_{\alpha\beta}, \tag{1}$$

where E is the dc-link voltage, and L_f and R_f are inductance and resistance of output filter, respectively. The two control model inputs, reference signal and disturbance, can be independently treated in a linear time-invariant (LTI) system. The final signal is the sum of individual responses, as shown bellow:

$$\vec{i}_{\alpha\beta} = \left(\frac{C_i \cdot G_{id}}{1 + C_i \cdot G_{id}}\right) \cdot \vec{r}_{\alpha\beta} - \left(\frac{1}{1 + C_i \cdot G_{id}}\right) \cdot \vec{p}_{\alpha\beta}. \tag{2}$$

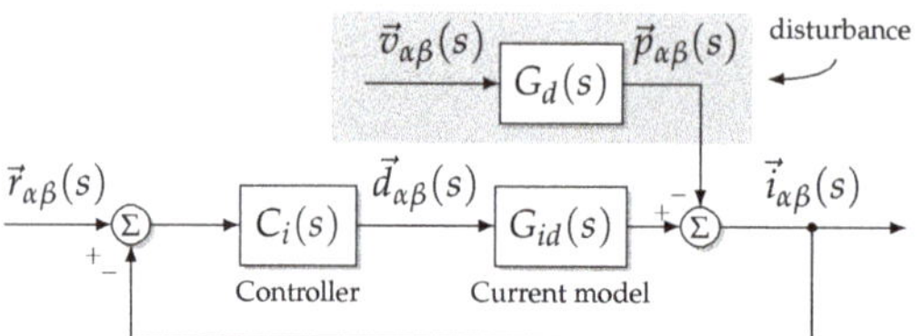

Figure 2. Simplified disturbance model for current control loop.

Therefore, in order to ensure reference signal tracking and disturbance rejection, condition (3) must be satisfied for all frequencies present in both the reference signal, which is generally a fundamental-frequency positive sequence signal and the disturbance, which is the grid voltage, containing typical harmonic components of positive and negative sequences:

$$\left|1 + C_i(j\omega) \cdot G_{id}(j\omega)\right| \gg 1. \tag{3}$$

Then, any disturbance will be rejected and the reference signal will be reproduced at the output:

$$\left|\frac{\vec{i}_{\alpha\beta}(j\omega)}{\vec{p}_{\alpha\beta}(j\omega)}\right| = \left|\frac{1}{1 + C_i(j\omega) \cdot G_{id}(j\omega)}\right| = M_p(\omega) \ll 1 \tag{4}$$

$$\left|\frac{\vec{i}_{\alpha\beta}(j\omega)}{\vec{r}_{\alpha\beta}(j\omega)}\right| = \left|\frac{C_i(j\omega) \cdot G_{id}(j\omega)}{1 + C_i(j\omega) \cdot G_{id}(j\omega)}\right| = M_r(\omega) = 1. \tag{5}$$

For a system subject to disturbances of different frequencies, such as in the case of voltage regulation when non-linear loads are connected to the PCC, disturbance rejection is an important design requirement.

The voltage v_{pcc} and current i_g of the electric network and the currents of the linear and nonlinear loads fed by a weak network are shown in Figure 3. The distorted current $i_{l,nl}$ absorbed by the nonlinear load distorts the PCC voltage and affects even the currents $i_{l,l}$ of the connected linear loads.

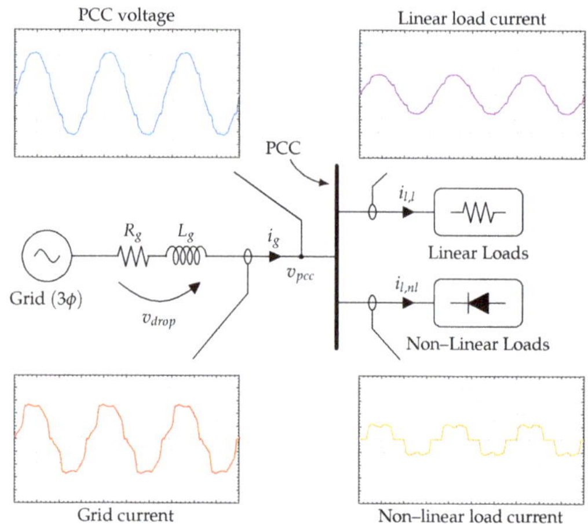

Figure 3. Effects of linear and non-linear loads fed by a weak grid.

2.1. Current Control Strategy

According to the internal model principle (IMP) the controller can reject disturbances or track references if it incorporates the disturbance model or the reference signal in its transfer function [30].

Thus, multiple resonant poles are required to guarantee null error at steady state if the reference or disturbance signal is distorted, i.e., have multiple harmonic frequencies.

In this sense, we used the SV-RC based on the inverse transfer function of the generalized delayed signal cancellation (iGDSC) [27] to control the converters' output current. Thus, a proper disturbance rejection capability can be added to the adopted voltage regulation scheme. Such SV-RC has poles located at multiple frequencies $j(n \cdot k + m)\omega_1, k \in \mathbb{Z}$ in the complex-plane and produces the desired performance with multiple harmonic components. The iGDSC SV-RC transfer function is shown in (6):

$$\vec{C}_{gdsc}(z) = \frac{1}{\vec{a}\left(1 - e^{j\frac{m}{n}2\pi} \cdot z^{-\frac{N}{n}}\right)}, \tag{6}$$

where the terms $\{n, m\} \in \mathbb{N}^*$ depend on the family of desired harmonic components. The coefficient N is the ratio of the sampling frequency to the fundamental frequency f_s/f_1. The operator $\vec{a}$ is a complex number and changes the controller amplitude and phase, however in this application the operator is chosen to maintain the unit gain without additional displacement to the controller, which leads to a real term.

The controller transfer function $\vec{C}_{gdsc}$ has the space-vector error signal $\vec{e}_{\alpha\beta}$ as input and the space-vector control action signal $\vec{u}_{\alpha\beta}$ as output, as shown in (7). Therefore, the final control action signal is obtained in (8) through the discrete transfer function.

$$\vec{C}_{gdsc}(z) = \frac{\vec{u}_{\alpha\beta}(z)}{\vec{e}_{\alpha\beta}(z)} \tag{7}$$

$$\vec{u}_{\alpha\beta}[k] = \frac{1}{a} \cdot \vec{e}_{\alpha\beta} + e^{j\frac{m}{n}2\pi} \cdot \vec{u}_{\alpha\beta}[k - N/n] \tag{8}$$

The SV-RC controller block diagram is shown in Figure 4. This structure has a positive feedback with a delay block $z^{-N/n}$, which produces periodic signal generator, multiplied by a complex value $e^{j(m/n)2\pi}$ in series with the delay block. The output signal $\vec{u}_{\alpha\beta}(z)$ is the sum of the positive feedback signal with the error signal $\vec{e}_{\alpha\beta}(z)$ multiplied by a complex value $1/\vec{a}$.

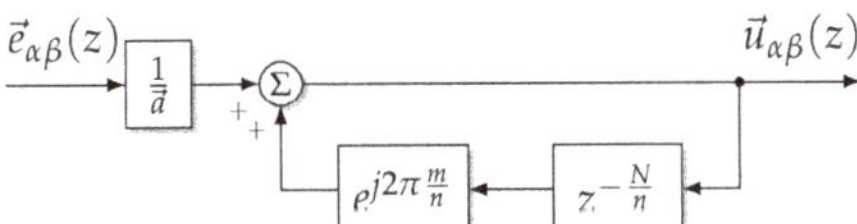

Figure 4. Block diagram of iGDSC based SV-RC with positive feedback structure.

As described in [27], the Nyquist diagram of the plant to be controlled must be inside the so called stability domain of the repetitive control scheme. Since the origin of the complex plane is not included in the stability domain of the iGDSC SV–RC, it is not possible to obtain adequate performance indexes to control the plant with the design. To increase the controller's stability domain, i.e., to enhance the system stability, a low-pass filter with unit gain in low frequencies and high attenuation rate in the high frequencies is used. The discrete implementation allows the use of an unconditionally stable digital finite impulse response (FIR) filter with linear phase delay when used with symmetric coefficients [31], as shown in $Q_M(z)$:

$$Q_M(z) = b_0 z^0 + \cdots + b_{(M-1)} z^{-(M-1)} + b_M z^{-M}, \tag{9}$$

where coefficients $b_0, \cdots, b_{(M-1)}, b_M$ are symmetric and M is an even number representing the FIR filter order.

The FIR filter can be represented by a constant magnitude part and a linear phase displacement part:

$$Q_M(j\omega) = \overbrace{\{\tilde{Q}(\omega)\}}^{\text{constant magnitude}} \cdot \underbrace{e^{-j\omega\frac{M}{2}}}_{\text{linear phase displacement}}. \tag{10}$$

Hence, the filter phase shift is known $(M/2)$ and can be cancelled by reducing the controller delay samples $(z^{-N/n})$ since this delay also produces a linear phase delay. Let $k_d = N/n$ be the original delay and k'_d the compensated delay:

$$k'_d = k_d - \frac{M}{2} = \frac{N}{n} - \frac{M}{2}. \tag{11}$$

The final transfer function is a result of the FIR filter and the linear phase compensation, as shown in (12):

$$\vec{C}_{igdsc}(z) = \frac{1}{\vec{a}} \cdot \frac{1}{1 - e^{j2\pi\frac{m}{n}} \cdot Q_M(z) \cdot z^{-k'_d}}. \tag{12}$$

In the digital implementation the computational delay is the time required by the processor to compute a new value from an input sample. This means that there is a delay with the modulation period in the control algorithm. This unit delay produces a response with a smaller phase margin and consequently the step response tends to be underdamped [32]. Therefore, the system may become unstable depending on the control design parameters.

A phase lead compensator is used to attenuate the undesired effect of computational delay. The compensator $H_l(w)$ has the transfer function defined by

$$H_l(w) = k_l \cdot \left(\frac{w + \omega_z}{w + \omega_z/k_f}\right), \quad \text{for } \omega_m = \frac{\omega_z}{\sqrt{k_f}}, \tag{13}$$

where k_l is the compensator gain, ω_z is the zero position, ω_z/k_f is the pole position, ω_m is the maximum lead frequency. The term k_f defines the distance between the compensator pole and zero. The condition $0 < k_f < 1$ must be satisfied to maintain the phase lead characteristic:

$$\sin(\varphi_m) = \frac{1 - k_f}{1 + k_f} \quad \therefore \quad k_f = \frac{1 - \sin(\varphi_m)}{1 + \sin(\varphi_m)}, \tag{14}$$

where φ_m is the maximum forward angle, with the following condition: $\varphi_m \leq 65°$ and $k_f \geq 0.05$ due to pole position physical limitation [33].

2.2. Design of the Current Controller

This loop must have fast dynamic response and high gain to allow proper reference tracking capability. In Figure 5, all the elements of the internal current control loop are shown: controller $C_i(z)$ designed based on the iGDSC operation, RC gain k_a, phase lead compensator $H_l(z)$, computational delay z^{-1} with the digital modulator DPWM, represented by $G_{\text{DPWM}}(w)$, transfer function $G_{id}(s)$ of the converter filter output current and current sensor $H_i(s)$.

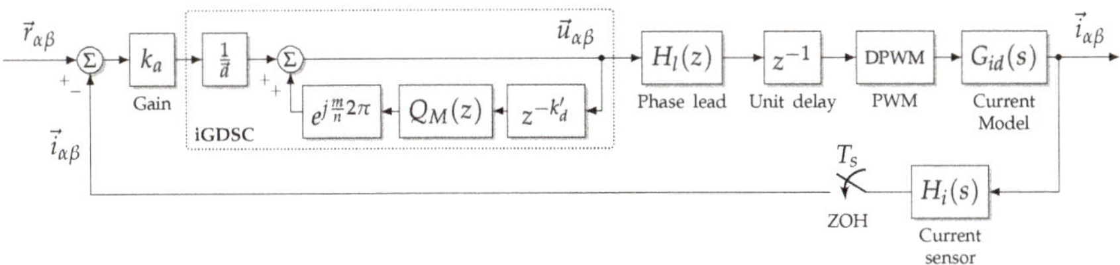

Figure 5. Block diagram of iGDSC based SV-RC with positive feedback structure.

In Figure 6 Bode (a) and Nyquist (b) diagrams of the output current model and controller open-loop transfer function are shown. The final open-loop transfer function resulted in a zero crossing frequency of 1.19 kHz, with 21.0° of phase margin (PM) and 6.04 dB of gain margin (GM). The Nyquist diagram is used to check the sensitivity function index value $\eta = 0.352$, which provides additional information about the controller stability, especially when exists multiple resonance peaks and multiple 0 dB crossings [34]. The sensitivity function is defined as the inverse of the distance between the Nyquist path and the critical point $(-1 + j0)$. The peak sensitivity is the maximum value of the sensitivity function, so higher values indicate the proximity to system instability [35].

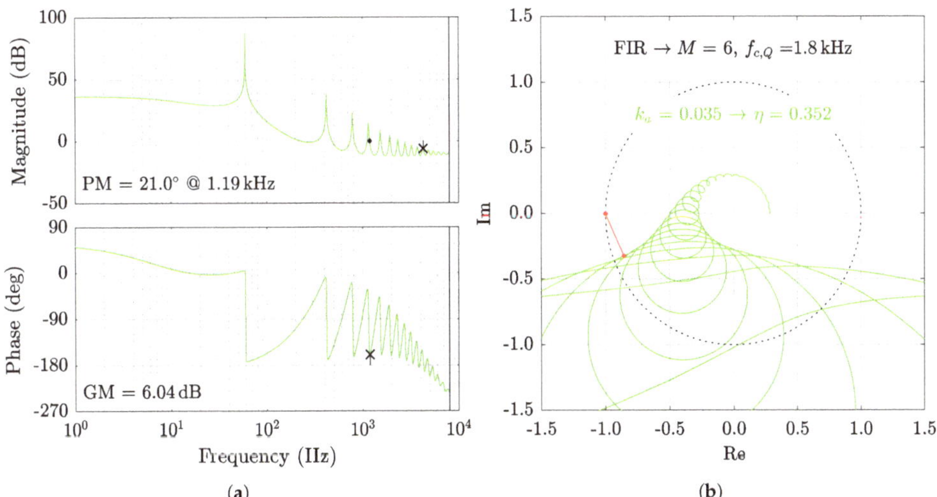

Figure 6. Nyquist and Bode diagram of final transfer function: (**a**) Bode diagram; (**b**) Nyquist diagram.

The cut-off frequency of the FIR filter of order $L = 6$ was set to 1.8 kHz which represents attenuation from the 30th harmonic component. The phase lead compensator was designed with maximum phase advance of $\phi_m = 33°$ at the frequency of $\omega_m = 2.2$ kHz to compensate the effects of computational delay, which results $k_f = 0.2948$ and $k_l = 1.0$. Moreover, the proportional gain needed to obtain the performance parameters previously presented was $k_a = 0.035$.

2.3. Effective Voltage Design Control

This loop controls PCC voltage as a function of the converter current. This control is done by an outer control loop with a slower dynamic than the inner current loop. The difference between loop dynamics has the purpose of decoupling outer and inner control loops and allowing their independent designs. The PCC voltage model is obtained from the equivalent circuit shown in

Figure 7. In this model, the converter current i_f is the controlled variable and the other variables (i_{load}, v_g) are considered disturbances for the voltage control loop.

$$v_{pcc}(s) = G_v(s) \cdot \left[i_f(s) - i_{load}(s) + v_g(s)/Z_g\right]. \tag{15}$$

Therefore, considering load current and grid voltage disturbances for $i_{load}(s) = 0$ and $v_g(s) = 0$:

$$G_v(s) = \frac{v_{pcc}(s)}{i_f(s)} = \frac{sL_g + R_g}{s^2 L_g C_f + sR_g C_f + 1} \tag{16}$$

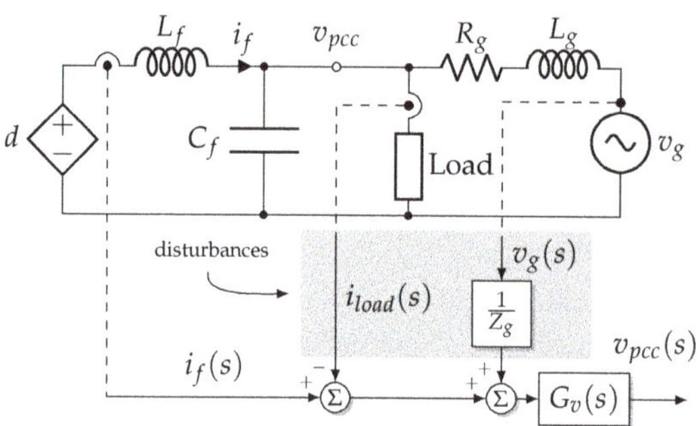

Figure 7. PCC voltage equivalent model with disturbances.

A proportional-integral (PI) controller with additional pole is used for outer loop control. This controller is suitable for constant signals since the integral action has null regime error for this input type. The extra pole is positioned at high frequency to attenuate the effects of higher order harmonic components. The transfer function is shown in (17):

$$C_v(s) = k_v \cdot \frac{s + z_v}{s(s + p_v)}. \tag{17}$$

where, k_v is the controller gain, z_v is the zero position and p_v is the pole position, both in rad/s.

The control design in the w-plane uses zero-order hold (ZOH) and Tustin discretization methods that allow a frequency response design with well known techniques of the s-plane, taking into account the effects of discretization [36].

The controller is designed with a cross-over frequency of 6 Hz and phase margin of 63°. The outer loop sampling frequency is adjusted to be ten times less $(f_s/10)$ than the inner loop. Thus the coefficients of the discrete transfer function preserve the precision required for the digital implementation.

3. Coordinated Power Control

As stated earlier, the use of reactive power alone to compensate voltage drop issues for weak LV grids is not as effective as it would be if it was applied to high-voltage systems. On the other hand, the active-power-based voltage regulation provides an improved performance, mainly for grids with high R/X ratio. However, such a technique would demand large energy storage devices if long intervals of operation were required. In this sense, the use of a coordinated control method of active and reactive power for voltage regulation is preferable especially when minimum use of the stored energy is required. Therefore, such feature allows the optimization of the ancillary service minimizing active power injection and, consequently, increasing the batteries' lifespan.

The active and reactive power coordinated strategy adopted in this paper was proposed by [15] and is performed by some logical operations, which are implemented along with the outer control loop as shown in Figure 8. In such scheme, the in-phase (i_{0°) and quadrature (i_{90°) components of the converter's reference current are provided in order to maintain the PCC voltage according to its desired range, controlling the amount of active power injected to the minimum required. Therefore, only severe voltage drops would demand active power injection to be compensated.

In order to avoid overlapping of both C_{0° and C_{90° control actions, the switch Sw_1 remains opened when the current i_{90° is lower than the converter rated output current, that is $i_{90^\circ} < 1$ (in pu). If such current component reaches the converted rated current without being able to restore the PCC voltage at its desired range, the switch Sw_1 is turned on. Thus, a small amount of the in-phase current component starts to be provided by the converter in order to enhance the voltage regulation. To hold the apparent power within the converter limits, the power limitation block imposes a dynamic saturation value $(i_{90^\circ}^{max})$ for the quadrature current i_{90°, which is equal to $\sqrt{1-i_{0^\circ}^2}$ (in pu). Once the desired voltage level is achieved, the amount of i_{0° and i_{90° are maintained until any other voltage deviation occur.

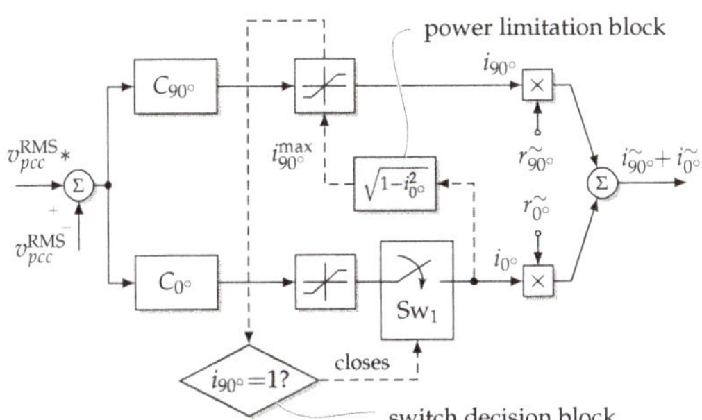

Figure 8. Logic algorithm used within outer control loop for voltage control.

3.1. *Effective Voltage Computation Method*

Aiming to guarantee a suitable operation, a precise acquisition of the grid voltage rms value is required. In [15] the authors use the measured rms voltage of only one phase of the system to control the voltage level of all three phases. Then, if the grid voltages are deeply unbalanced, such adoption may lead to inaccurate results and even to severe overvoltage condition to the other two phases that are not considered. In this regard, we have used the effective voltage as a measure of the grid rms voltage which is appropriate for three-phase systems since it aims to take into account the rms voltage levels of the three phases simultaneously.

The phase-to-phase effective voltage for a three-wire system is defined as [37]:

$$v_e = \sqrt{\frac{v_{ab}^2 + v_{bc}^2 + v_{ca}^2}{3}} \tag{18}$$

The phase-to-phase voltages (v_{ab}, v_{bc}) are used to obtain the Clarke's transformation components (v_α, v_β):

$$\begin{bmatrix} v_\alpha \\ v_\beta \end{bmatrix} = \begin{bmatrix} 2/3 & 1/3 \\ 0 & \sqrt{3}/3 \end{bmatrix} \cdot \begin{bmatrix} v_{ab} \\ v_{bc} \end{bmatrix} \tag{19}$$

The phase-to-phase effective voltage is calculated using v_α and v_β from (19):

$$v_{ef} = \sqrt{3} \cdot \sqrt{\frac{v_\alpha^2 + v_\beta^2}{2}} \tag{20}$$

However, the network voltages may have several harmonic components that can lead to distorted values of effective voltage and this distortion becomes relevant in weak grids. For that reason we have adopted the generalized delayed signal cancellation (GDSC) method to detect the fundamental frequency positive sequence (FFPS) of voltage signal [38].

This transformation consists of a complex mathematical operation on a space vector. Current and delayed samples of the input space vector are used to cancel a particular harmonic component family. The GDSC operation is described as

$$\vec{f}_{\text{GDSC}}(z) = \underbrace{\vec{a} \cdot [1 - e^{j\frac{m}{n}2\pi} \cdot z^{-\frac{N}{n}}]}_{\vec{G}_{\text{GDSC}}} \cdot \vec{s}_{\alpha\beta}(z). \tag{21}$$

where the terms of the set $n \cdot k + m, \forall k \in \mathbb{Z}$ and $n > m \geq 0$ define the harmonic component family, n defines the repetition period, m defines the initial frequency component, $\vec{a}$ is the complex operator that defines the operation gain, $\vec{s}_\alpha$ is the input space vector and $\vec{f}_{\text{GDSC}}$ is the output space vector.

It is not possible to eliminate all the harmonic components to obtain the FFPS signal vector with only one GDSC transformation, so five cascaded operations are designed to eliminate most of the harmonic components. Each operation eliminates a certain family of harmonic components and when cascaded allows the extraction of FFPS [38].

The first designed transformation eliminates the $2k + 2$ family, the second transformation eliminates the $4k + 3$ family, the third transformation eliminates the $8k + 5$ family, the fourth transformation eliminates the $16k + 9$ family, the fifth transformation eliminates the family $32k + 17$, leaving only the components of the set $\{\cdots, -31, +1, +33, \cdots\}$.

In Figure 9 the filter frequency response magnitude graph is shown. In this graphic, the FFPS has unitary amplitude, i.e., the filter allows FFPS to pass and eliminates or attenuates almost completely the other components. Such components that still persist in the bandwidth are of high order and are practically nonexistent in the measurement system and do not interfere with the final response.

Therefore, the effective voltage is then called as effective positive-sequence voltage after the FFPS-GDSC filter operation:

$$v_{ef}^{+1} = \sqrt{3} \cdot \sqrt{\frac{\left(v_\alpha^{+1}\right)^2 + \left(v_\beta^{+1}\right)^2}{2}} \tag{22}$$

where, v_α^{+1} and v_β^{+1} are the resulting voltage components from $\alpha\beta$ transformation after FFPS-GDSC operation.

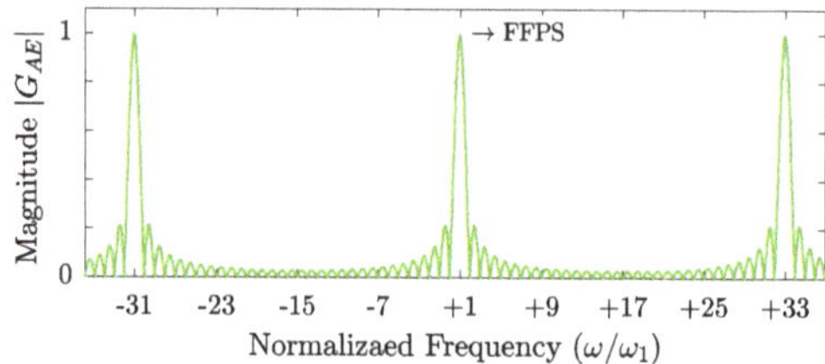

Figure 9. Magnitude graph of operation FFPS-GDSC as a function of normalized frequency.

3.2. Active and Reactive Power Reading Method

The power calculation uses the voltage and current components in $\alpha\beta$ coordinates. This information is used to check the operating point and the voltage regulation strategy, then it is not used for control or reference generation.

The voltage components filtered by the FFPS-GDSC operation are used to calculate the active and reactive power in three-phase systems:

$$\begin{bmatrix} p \\ q \end{bmatrix} = \frac{3}{2} \cdot \begin{bmatrix} v_\alpha^{+1} & v_\beta^{+1} \\ -v_\beta^{+1} & v_\alpha^{+1} \end{bmatrix} \cdot \begin{bmatrix} i_\alpha \\ i_\beta \end{bmatrix} \tag{23}$$

The abc-$\alpha\beta$ blocks of coordinate transformations are shown in Figure 10, they are: the FFPS-GDSC filter block, power and effective voltage calculation.

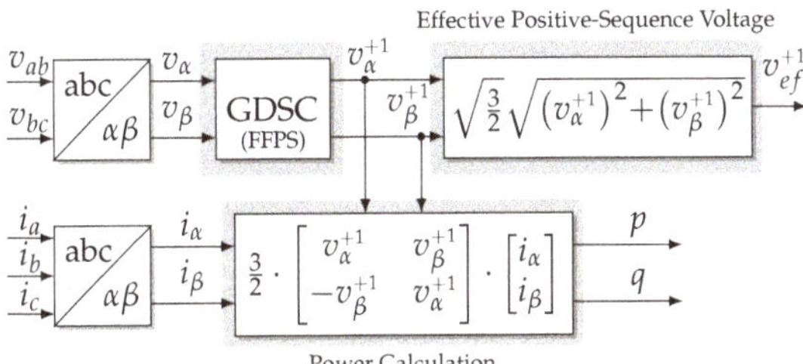

Figure 10. Method for computing the effective voltage and the active and reactive power calculation.

4. Experimental Results and Discussions

A three-phase three-wire 3.8 kVA prototype was built in order to demonstrate the operation of the proposed control algorithm, as shown in Figure 11. The converter was connected to a 127 V/60 Hz grid in series with a impedance (Z_g) formed by a 3.10 Ω resistor and a 3.80 mH inductor, that results in 2.16 of R/X impedance ratio, which is compatible with a weak distribution grid. The converter consists of a three-phase voltage source inverter (VSI) with a switching frequency of 18 kHz, with a second order output filter with 3.50 mH inductance and 5.0 μF capacitance for switching frequency ripple attenuation. The DC-link of 6600 μF capacitance operates with 500 V of DC voltage. The power modules are mounted on a forced-ventilation aluminum heater and drivers with isolated control signals.

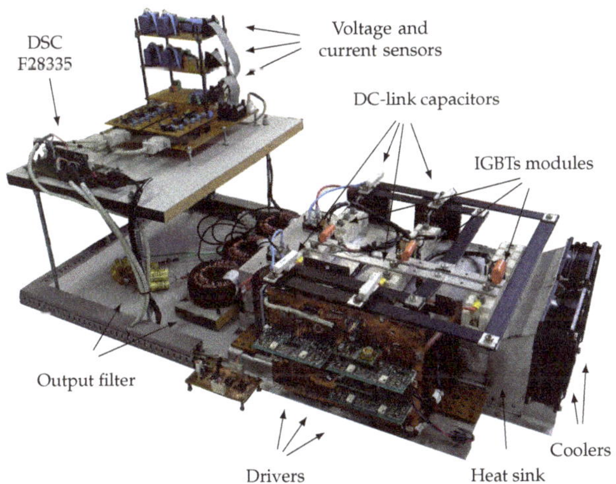

Figure 11. Three-phase prototype built in laboratory.

It is worth to mention that the currents and voltages measuring tips were attached in the points stated in Figure 7 for all experimental tests that were developed. Thus the converter currents (i_{fa}, i_{fb} and i_{fc}) were measured in the output filter inductors of each phase and the voltage $v_{pcc,ab}$ is the phase-to-phase voltage of the converter output filter capacitor. Such quantities measurements were acquired in the high resolution acquisition mode of the oscilloscope and no internal filter was used.

The setup for experimental results is shown in Figure 12. Converter currents and PCC phase-to-phase voltage in steady state operation are shown in Figure 13a. For comparison purposes, the voltage before the regulation is shown in R2 (black) and after the regulation is shown in CH4 (green). In this test, linear resistive loads of 56 Ω and a three-phase Graetz bridge rectifier are used as linear and nonlinear loads, with commutation inductor of 560 µH and resistive dc load of 40.67 Ω as shown in Figure 12a.

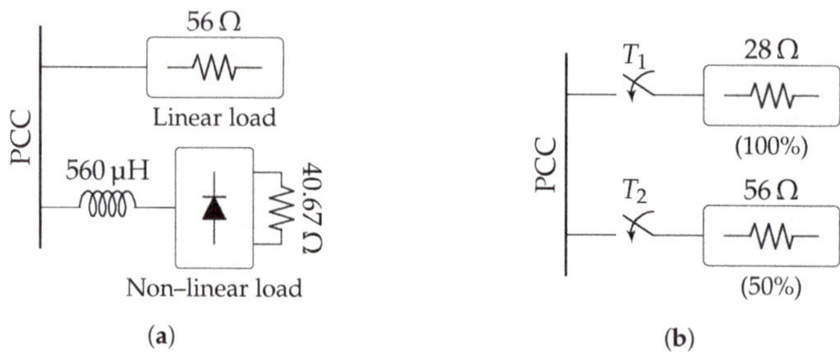

(**a**) (**b**)

Figure 12. Setup for experimental results: (**a**) Steady state results with non-linear loads; (**b**) Linear load step variation.

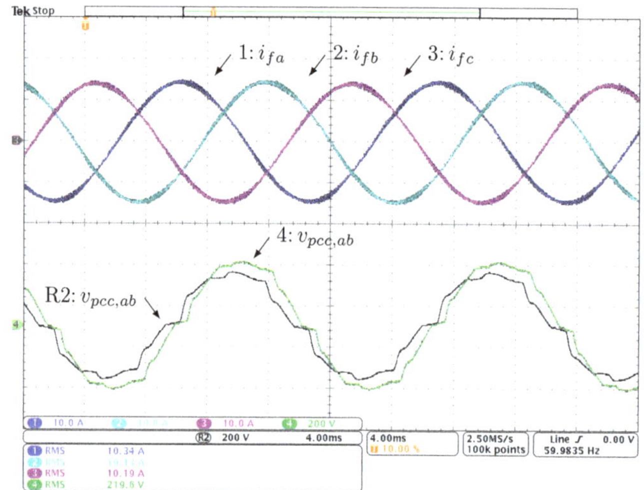

(**a**) CH1: i_{fa} [10 A/div], CH2: i_{fb} [10 A/div], CH3: i_{fc} [10 A/div], CH4: $v_{pcc,ab}$ [200 V/div], R2: $v_{pcc,ab}$ [200 V/div]

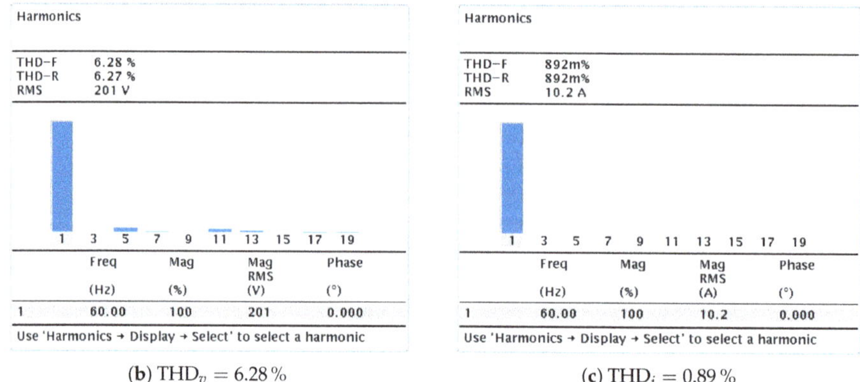

(**b**) $\text{THD}_v = 6.28\,\%$ (**c**) $\text{THD}_i = 0.89\,\%$

Figure 13. Experimental results in steady state: (**a**) Converter currents with low harmonic distortion even with grid distorted voltage; (**b**) Harmonic voltage analysis; (**c**) Harmonic current analysis.

The voltage harmonic distortion is the result of a nonlinear load connected in a weak network. It is possible to observe the increase of effective voltage value to nominal value of 220 V with a current of 10 A in each phase with both active and reactive power compensation. Voltage and current total harmonic distortion (THD) are shown in Figure 13b,c. Voltage THD_v is 6.28% and current THD_i is 0.89%, which shows that the controller can reject the disturbances from the high order harmonic components and ensures a pure sinusoidal current.

Steady state waveforms for linear loads are shown in Figure 14a. The currents injected by the converter are composed of 0.92 pu quadrature and 0.39 pu in-phase components resulting in 10 A to regulate the PCC effective phase-to-phase voltage at its nominal value of 220 V. Transient results are shown in Figure 14b at the moment the converter starts its operation with the 28 Ω load. It is possible to check the voltage regulation from 193 V to 220 V, while the currents i_{90° and i_{0° are changed by the control algorithm. The quadrature reference current reaches its maximum value of 1 pu and from that moment on, the in-phase reference current is activated to reach the desired voltage regulation. Reduction of the quadrature reference current to 0.92 pu is necessary to keep the converter within the maximum power specification. Finally, the active and reactive power graphs shows power injection of 3450 VAr and 800 W.

The results for load step variation are shown in Figure 15a In this test, two loads are selected: 100% (28 Ω) and 50% (56 Ω) of nominal load as shown in Figure 12b Initially, the 50% load is connected, being replaced by 100% load. In first test, the 50% load requires only reactive power of 2700 VAr for voltage regulation, however the active power injection (800 W) becomes necessary when the load increases to the nominal values, which can be seen in Figure 15b. The operation duality of the technique works in similar manner, thus, the load variation occurs in the reverse direction, i.e., the load is reduced. As expected, the control returns to the reactive power-only injection condition, which correspond to 0.70 pu of quadrature reference.

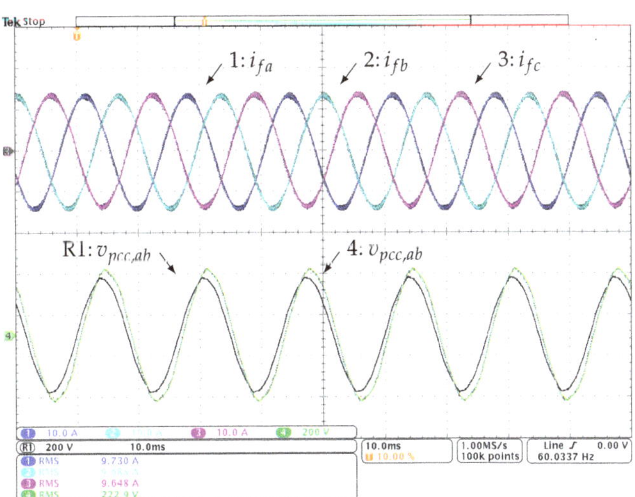

(**a**) CH1: i_{fa} [10 A/div], CH2: i_{fb} [10 A/div], CH3: i_{fc} [10 A/div], CH4: $v_{pcc,ab}$ [200 V/div], R2: $v_{pcc,ab}$ [200 V/div]

Figure 14. *Cont.*

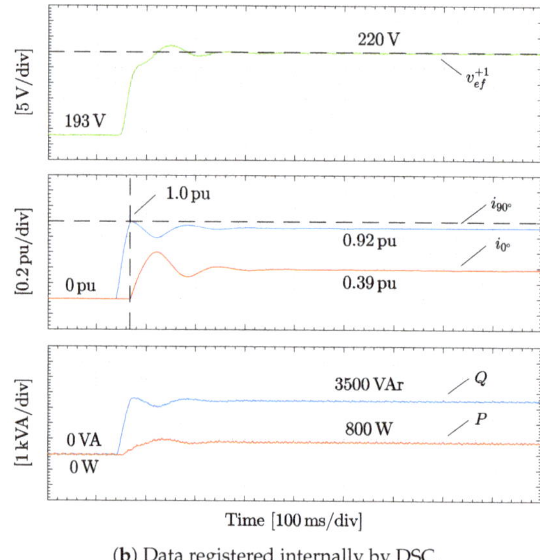

(**b**) Data registered internally by DSC.

Figure 14. Experimental results: (**a**) Converter currents for voltage regulation with linear loads; (**b**) Results for transient regime: effective voltage, in-phase and quadrature current references and converter injected power.

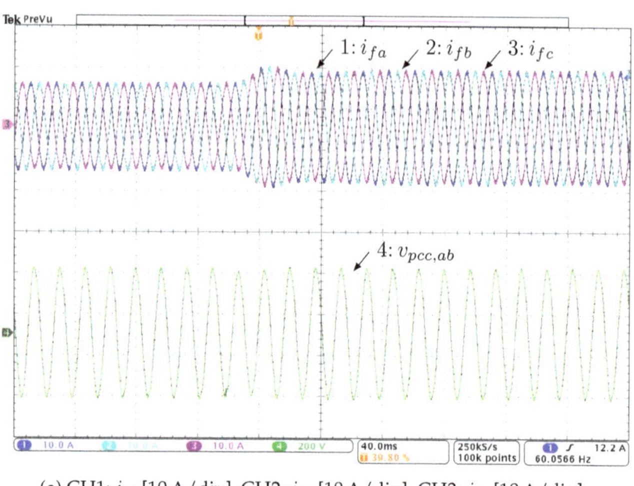

(**a**) CH1: i_{fa} [10 A/div], CH2: i_{fb} [10 A/div], CH3: i_{fc} [10 A/div], CH4: $v_{pcc,ab}$ [200 V/div]

Figure 15. *Cont.*

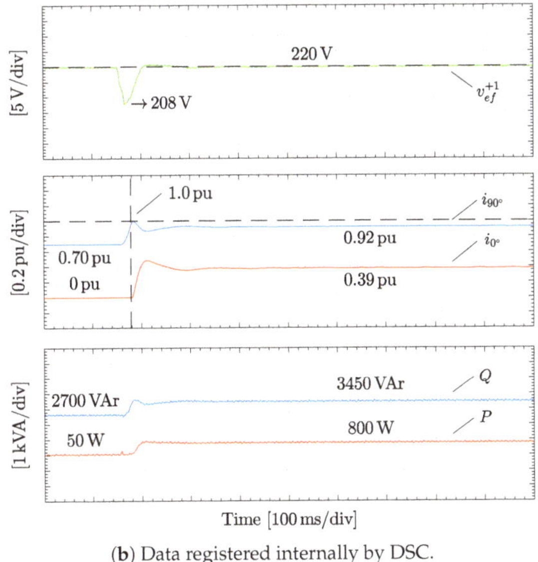

(**b**) Data registered internally by DSC.

Figure 15. Experimental results for load change: (**a**) Converter currents and PCC voltage for load increase; (**b**) Waveform of positive sequence effective voltage, quadrature and phase current references, and converter injected power to load increase.

5. Conclusions

In this paper, we proposed a voltage regulation scheme with an enhanced disturbance rejection capability for LV grids using repetitive vector-control. Once voltage regulation in high R/X ratio networks has proved to be more influenced by active power rather than reactive power, we have used a coordinated control based on active and reactive power injection to compensate voltage drops. In this sense, such control approach may improve lifespan and autonomy of energy storage systems when compared to conventional active power injection approaches, once it is implemented to establish priority of reactive power over active power injection. Thus, the active power stored is only used for severe voltage drops.

The SV-RC we have used may enhance both disturbance rejection and precise reference tracking capabilities for the converter output current imposition, which is desirable for weak grids once it is more susceptible to harmonic voltage distortion caused by non-linear load currents. Additionally, we used the three-phase collective rms value to implement the control method, instead of one phase voltage rms value. Then, the risk of erroneous voltage regulation due to unbalance or single-phase voltage sag/swell is avoided.

Through experimental results, we can note that the proposed control scheme proves to be more effective than previous ones, ensuring low-distorted in all three phases, even when highly distorted grid voltage appears at the PCC. Besides, while the dynamic response has shown to be very fast, the steady-state response has been extremely precise, once the proposed scheme has been able to achieve the desired rms voltage reference value in few cycles when subjected to a load step.

Finally, we believe that the voltage regulation approach proposed in this paper should be very useful to be implemented within EV charging stations that have been built far from large power stations. In this regard, the energy stored in EVs batteries can be used as an ancillary service in order to compensate severe voltage drops, without significantly compromising lifespan of energy storage devices.

Author Contributions: Validation, visualization, F.J.Z.; formal analysis, methodology, writing–original draft preparation, F.J.Z. and E.V.S.; funding acquisition, resources, F.A.S.N., A.L.B. and M.M.; conceptualization, investigation, writing–review and editing, F.J.Z., E.V.S, F.A.S.N., A.L.B. and M.M. All authors have read and agreed to the published version of the manuscript.

Funding: This work was developed with the support of the Programa Nacional de Cooperação Acadêmica da Coordenação de Aperfeiçoamento de Pessoal de Nível Superior—CAPES/Brazil—Finance Code 001. This research was funded by Conselho Nacional de Desenvolvimento Científico e Tecnológico (CNPq) grant number 465640/2014-1, Coordenação de Aperfeiçoamento de Pessoal de Nível Superior (CAPES) grant number 23038.000776/2017-54, Fundação de Amparo à Pesquisa do Estado do Rio Grande do Sul (FAPERGS) grant number 17/2551-0000517-1, Fundação de Amparo à Pesquisa e Inovação do Estado de Santa Catarina (FAPESC), Instituto Nacional de Ciência e Tecnologia em Geração Distribuída de Energia Elétrica (INCT–GD), Fundação de Amparo à Ciência e Tecnologia do Estado de Pernambuco (FACEPE), Universidade Federal de Pernambuco (UFPE) and Universidade do Estado de Santa Catarina (UDESC).

Conflicts of Interest: The authors declare no conflict of interest.

Abbreviations

The following abbreviations are used in this manuscript:

DPWM	Digital pulse-width modulation
DSTATCOM	Distribution static synchronous compensator
ESS	Energy storage system
FFPS	Positive sequence fundamental frequency
FIR	Finite impulse response
GM	Gain margin
iGDSC	Inverse transfer function of the generalized delayed signal cancellation
IMP	Internal model principle
LC	Low-voltage
LTI	Linear time-invariant
PCC	Point of common connection
PI	Proportional-integral
PM	Phase margin
PR	Proportional-resonant
PWM	Pulse-width modulation
RC	Repetitive control
SV-RC	Space-vector repetitive controller
THD	Total harmonic distortion
VSI	Voltage source inverter
ZOH	Zero-order hold

References

1. Wang, Y.; Tan, K.T.; Peng, X.Y.; So, P.L. Coordinated Control of Distributed Energy-Storage Systems for Voltage Regulation in Distribution Networks. *IEEE Trans. Power Deliv.* **2016**, *31*, 1132–1141. [CrossRef]
2. Shu, D.; Xie, X.; Rao, H.; Gao, X.; Jiang, Q.; Huang, Y. Sub- and Super-Synchronous Interactions Between STATCOMs and Weak AC/DC Transmissions With Series Compensations. *IEEE Trans. Power Electron.* **2018**, *33*, 7424–7437. [CrossRef]
3. Kumar, C.; Mishra, M.K. An Improved Hybrid DSTATCOM Topology to Compensate Reactive and Nonlinear Loads. *IEEE Trans. Ind. Electron.* **2014**, *61*, 6517–6527. [CrossRef]
4. Hu, J.; Li, Z.; Zhu, J.; Guerrero, J.M. Voltage Stabilization: A Critical Step Toward High Photovoltaic Penetration. *IEEE Ind. Electron. Mag.* **2019**, *13*, 17–30. [CrossRef]
5. Marra, F.; Yang, G.Y.; Træholt, C.; Larsen, E.; Østergaard, J.; Blažič, B.; Deprez, W. EV Charging Facilities and Their Application in LV Feeders With Photovoltaics. *IEEE Trans. Smart Grid* **2013**, *4*, 1533–1540. [CrossRef]
6. Kabir, M.N.; Mishra, Y.; Ledwich, G.; Dong, Z.Y.; Wong, K.P. Coordinated Control of Grid-Connected Photovoltaic Reactive Power and Battery Energy Storage Systems to Improve the Voltage Profile of a Residential Distribution Feeder. *IEEE Trans. Ind. Inform.* **2014**, *10*, 967–977. [CrossRef]

7. Carvalho, P.M.S.; Correia, P.F.; Ferreira, L.A.F.M. Distributed Reactive Power Generation Control for Voltage Rise Mitigation in Distribution Networks. *IEEE Trans. Power Syst.* **2008**, *23*, 766–772. [CrossRef]
8. Teshome, D.F.; Xu, W.; Bagheri, P.; Nassif, A.B.; Zhou, Y. A Reactive Power Control Scheme for DER-caused Voltage Rise Mitigation in Secondary Systems. *IEEE Trans. Sustain. Energy* **2018**, *10*, 1684–1695. [CrossRef]
9. Mazza, A.; Mirtaheri, H.; Chicco, G.; Russo, A.; Fantino, M. Location and Sizing of Battery Energy Storage Units in Low Voltage Distribution Networks. *Energies* **2019**, *13*, 52. [CrossRef]
10. Alshehri, J.; Khalid, M.; Alzahrani, A. An Intelligent Battery Energy Storage-Based Controller for Power Quality Improvement in Microgrids. *Energies* **2019**, *12*, 2112. [CrossRef]
11. Alzahrani, A.; Alharthi, H.; Khalid, M. Minimization of Power Losses through Optimal Battery Placement in a Distributed Network with High Penetration of Photovoltaics. *Energies* **2019**, *13*, 140. [CrossRef]
12. Torres-Moreno, J.; Gimenez-Fernandez, A.; Perez-Garcia, M.; Rodriguez, F. Energy Management Strategy for Micro-Grids with PV-Battery Systems and Electric Vehicles. *Energies* **2018**, *11*, 522. [CrossRef]
13. Omar, N.; Monem, M.A.; Firouz, Y.; Salminen, J.; Smekens, J.; Hegazy, O.; Gaulous, H.; Mulder, G.; den Bossche, P.V.; Coosemans, T.; et al. Lithium iron phosphate based battery—Assessment of the aging parameters and development of cycle life model. *Appl. Energy* **2014**, *113*, 1575–1585. [CrossRef]
14. Wang, L.; Bai, F.; Yan, R.; Saha, T.K. Real-Time Coordinated Voltage Control of PV Inverters and Energy Storage for Weak Networks With High PV Penetration. *IEEE Trans. Power Syst.* **2018**, *33*, 3383–3395. [CrossRef]
15. Zimann, F.J.; Batschauer, A.L.; Mezaroba, M.; Neves, F.A.S. Energy storage system control algorithm for voltage regulation with active and reactive power injection in low-voltage distribution network. *Electr. Power Syst. Res.* **2019**, *174*, 105825. [CrossRef]
16. Cupertino, A.F.; Farias, J.V.M.; Pereira, H.A.; Seleme, S.I.; Teodorescu, R. Comparison of DSCC and SDBC Modular Multilevel Converters for STATCOM Application During Negative Sequence Compensation. *IEEE Trans. Ind. Electron.* **2019**, *66*, 2302–2312. [CrossRef]
17. Da Cunha, J.; Hock, R.; Oliveira, S.G.; Michels, L.; Mezaroba, M. A novel control scheme to reduce the reactive power processed by a Multifunctional Voltage-Quality Regulator. *Electr. Power Syst. Res.* **2018**, *163*, 348–355. [CrossRef]
18. Chen, D.; Zhang, J.; Qian, Z. Research on fast transient and $6n \pm 1$ harmonics suppressing repetitive control scheme for three-phase grid-connected inverters. *IET Power Electron.* **2013**, *6*, 601–610. [CrossRef]
19. Yang, Y.; Zhou, K.; Wang, H.; Blaabjerg, F. Analysis and Mitigation of Dead-Time Harmonics in the Single-Phase Full-Bridge PWM Converter With Repetitive Controllers. *IEEE Trans. Ind. Appl.* **2018**, *54*, 5343–5354. [CrossRef]
20. Cho, Y.; Lai, J. Digital Plug-In Repetitive Controller for Single-Phase Bridgeless PFC Converters. *IEEE Trans. Power Electron.* **2013**, *28*, 165–175. [CrossRef]
21. Zhou, K.; Wang, D. Digital repetitive learning controller for three-phase CVCF PWM inverter. *IEEE Trans. Ind. Electron.* **2001**, *48*, 820–830. [CrossRef]
22. Costa-Castelló, R.; Griñó, R.; Fossas, E. Odd-harmonic digital repetitive control of a single-phase current active filter. *IEEE Trans. Power Electron.* **2004**, *19*, 1060–1068. [CrossRef]
23. Zhou, K.; Low, K.S.; Wang, D.; Luo, F.L.; Zhang, B.; Wang, Y. Zero-phase odd-harmonic repetitive controller for a single-phase PWM inverter. *IEEE Trans. Power Electron.* **2006**, *21*, 193–201. [CrossRef]
24. Escobar, G.; Hernandez-Briones, P.G.; Martinez, P.R.; Hernandez-Gomez, M.; Torres-Olguin, R.E. A Repetitive-Based Controller for the Compensation of $6\ell \pm 1$ Harmonic Components. *IEEE Trans. Ind. Electron.* **2008**, *55*, 3150–3158. [CrossRef]
25. Lu, W.; Zhou, K.; Wang, D.; Cheng, M. A General Parallel Structure Repetitive Control Scheme for Multiphase DC–AC PWM Converters. *IEEE Trans. Power Electron.* **2013**, *28*, 3980–3987. [CrossRef]
26. Luo, Z.; Su, M.; Yang, J.; Sun, Y.; Hou, X.; Guerrero, J.M. A Repetitive Control Scheme Aimed at Compensating the $6k + 1$ Harmonics for a Three-Phase Hybrid Active Filter. *Energies* **2016**, *9*, 787. [CrossRef]
27. Zimann, F.J.; Neto, R.C.; Neves, F.A.S.; de Souza, H.E.P.; Batschauer, A.L.; Rech, C. A Complex Repetitive Controller Based on the Generalized Delayed Signal Cancelation Method. *IEEE Trans. Ind. Electron.* **2019**, *66*, 2857–2867. [CrossRef]
28. Clarke, E. *Circuit Analysis of A-C Power Systems*; J. Wiley & Sons: Hoboken, NJ, USA, 1943.
29. Mottola, F.; Proto, D.; Varilone, P.; Verde, P. Planning of Distributed Energy Storage Systems in μGrids Accounting for Voltage Dips. *Energies* **2020**, *13*, 401. [CrossRef]

30. Francis, B.; Wonham, W. The internal model principle of control theory. *Automatica* **1976**, *12*, 457–465. [CrossRef]
31. Escobar, G.; Mattavelli, P.; Hernandez-Gomez, M.; Martinez-Rodriguez, P.R. Filters with Linear-Phase Properties for Repetitive Feedback. *IEEE Trans. Ind. Electron.* **2014**, *61*, 405–413. [CrossRef]
32. Buso, S.; Mattavelli, P. *Digital Control in Power Electronics*; Lectures on Power Electronics; Morgan & Claypool Publishers: San Rafael, CA, USA, 2006; p. 151.
33. Ogata, K. *Modern Control Engineering*, 5th ed.; Prentice Hall: Upper Saddle River, NJ, USA, 2010.
34. Neto, R.C.; Neves, F.A.S.; de Souza, H.E.P.; Zimann, F.J.; Batschauer, A.L. Design of Repetitive Controllers Through Sensitivity Function. In Proceedings of the 2018 IEEE 27th International Symposium on Industrial Electronics (ISIE), Cairns, Australia, 13–15 June 2018; pp. 495–501. [CrossRef]
35. Yepes, A.G.; Freijedo, F.D.; Lopez, Ó.; Doval-Gandoy, J. Analysis and Design of Resonant Current Controllers for Voltage-Source Converters by Means of Nyquist Diagrams and Sensitivity Function. *IEEE Trans. Ind. Electron.* **2011**, *58*, 5231–5250. [CrossRef]
36. Ogata, K. *Discrete-time Control Systems*, 2nd ed.; Prentice Hall: Upper Saddle River, NJ, USA, 1995.
37. Sutikno, T. *IEEE Trial-Use Standard Definitions for the Measurement of Electric Power Quantities Under Sinusoidal, Non-sinusoidal, Balanced, Or Unbalanced Conditions*; IEEE Std 1459-2000; IEEE: Piscataway Township, NJ, USA, 2000; pp. 1–44. [CrossRef]
38. Neves, F.A.S.; Cavalcanti, M.C.; de Souza, H.E.P.; Bradaschia, F.; Bueno, E.J.; Rizo, M. A Generalized Delayed Signal Cancellation Method for Detecting Fundamental-Frequency Positive-Sequence Three-Phase Signals. *IEEE Trans. Power Del.* **2010**, *25*, 1816–1825. [CrossRef]

Article

Equivalent Phase Current Harmonic Elimination in Quadruple Three-Phase Drives Based on Carrier Phase Shift Method

Xuchen Wang [1], Hao Yan [2], Giampaolo Buticchi [1,*], Chunyang Gu [1] and He Zhang [1]

1 Key Laboratory of More Electric Aircraft Technology of Zhejiang Province, Department of electrical and electronic engineering, University of Nottingham Ningbo China, Ningbo 315100, China; xuchen.wang@nottingham.edu.cn (X.W.); chunyang.gu@nottingham.edu.cn (C.G.); he.zhang@nottingham.edu.cn (H.Z.)

2 School of Electrical and Electronic Engineering, Nanyang Technological University, Singapore 639798, Singapore; hao.yan@ntu.edu.sg

* Correspondence: giampaolo.buticchi@nottingham.edu.cn; Tel.: +86-(0)-574-881-800-00-8716

Received: 24 April 2020; Accepted: 22 May 2020; Published: 28 May 2020

Abstract: Multiphase drives are entering the spotlight of the research community for transportation applications with their high power density and the possibility of high fault tolerance. The multi three-phase drive is one of the main types of multiphase drives that allows for the direct adoption of commercial three-phase converters and high control flexibility. The elimination of high-frequency current harmonics will reduce the flux linkage harmonics, torque ripple, vibration and noise in machine drives. Therefore, this work introduces a new method to the modelling of equivalent phase current in multi three-phase drives with the double integral Fourier analysis method. A new carrier-based pulse-width modulation (CPWM) method is introduced to reduce the equivalent phase current harmonics by applying proper carrier phase angle to each subsystem in the multi three-phase drives. The proposed angles of carrier signals are analyzed for quadruple three-phase drives, and the corresponding experimental results confirm the significance of the proposed phase-shifted CPWM method to eliminate the equivalent phase current harmonics.

Keywords: multiphase drives; pulse width modulation; current harmonics

1. Introduction

Three-phase machines were universally used at the beginning of the 20th century as they present better torque performance compared with the single-phase and the two-phase machines, which produce the twice pulsating torque ripple [1]. The increasing phase number of the machine will not produce the twice-pulsating torque ripple. The development of power electronics in the 1980s enabled the adoption of machines with more than three phases. The machine can be connected to the power converters rather than directly connected to the three-phase power supply. Therefore, an arbitrary phase machine can be used as long as the phase number of the machine matches the phase number of the power converter, especially for the application of electric propulsion [2–5]. There are several reasons why the multiphase machine attracts a wide range of interest in the research community. Firstly, the multiphase machine has good fault tolerance performance compared with the conventional three-phase machine [6,7]. Taking a 15-phase machine as an example, the machine can be operated at over 90% of the rated power with the breakdown of one phase. Secondly, the adoption of a multiphase machine greatly reduces the voltage sharing on power converters of each phase leg, which offers the feasibility of the multiphase machine used for high-power applications [8–12]. Additionally, the control of a multiphase drive system becomes more flexible, which gives a possibility to cancel the time-domain harmonics and space-domain harmonics in the drive system. Therefore, the efficiency and torque performance in the multiphase is improved with respect to the traditional single three-phase drives [13–15].

Recently, multi three-phase machines with separate neutral points controlled by parallel converters have become an increasing concern, which is shown in Figure 1 [8,16]. The adoption of parallel converters make it possible to control very high-power and high-speed machines with commercial products, since the power requirement on each inverter module is reduced. Moreover, this topology has high fault tolerance, as any inverter module is independent without any electrical connections to the other modules [6]. Besides, the adoption of the parallel converters increases the flexibility of the control algorithm. The torque ripple, noise and vibration of the machine, and the direct current (dc) link voltage ripple are possibly to be reduced by applying a proper control scheme [17,18].

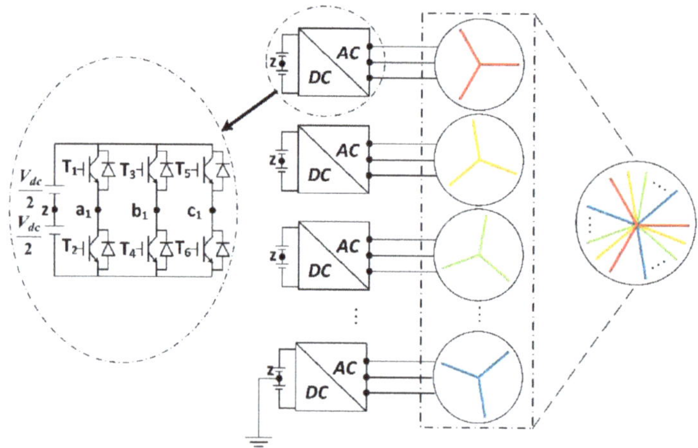

Figure 1. Multi three-phase drives system.

The carrier-based pulse width modulation (CPWM) is known as the earliest and most straightforward modulation technique [19]. The working principle of the CPWM is to compare the fundamental signal with the carrier signal (high frequency) to determine the states of switching devices on power converters. When the fundamental waveform is lower than the carrier with waveform, the switch at the bottom of the converter leg (T_2 T_4 and T_6 in Figure 1) will turn on; when the fundamental waveform is larger than the carrier waveform, the switch on the top of the converter leg (T_1 T_3 and T_5 in Figure 1) will turn on. Compared with the space vector modulation, the CPWM has lower dc link voltage utilization [20]. This drawback can be avoided by the injection of the zero-sequence harmonics [21]. The implementation of CPWM is straightforward and it can be extended to the control of the multiphase multilevel converters without greatly increasing the complexity of the control algorithm. The phase-shifted CPWM is the commonly used CPWM for the control of multilevel converters, as it has the same switching frequency and conducting periods, which means even power distribution on the power switches of the converter [22,23]. Recent research works have verified that the pulse width modulation (PWM) technique results in high-frequency torque ripple in multiphase drives [24,25]. For instance, [25] shows that the high-frequency torque harmonics are mitigated by choosing appropriate carrier angles in dual three-phase drive systems. The high-frequency PWM-related current harmonics are known as the source of the high-frequency flux linkage harmonics, torque ripple, vibration and noise [26,27]. Therefore, it is of importance to reduce the equivalent current harmonics of the multiphase drive systems by considering the effect the CPWM technique.

This article exploits the degrees of control freedom in multi three-phase drives, by applying the concept phase-shifted CPWM to multi three-phase drives to reduce total equivalent current harmonics.

The remaining part of the article includes the following sections: Section 2 introduces the double Fourier integral analysis method of pulse width modulation. Section 3 presents the mathematical modelling of the equivalent current harmonics in multi three-phase drives. Section 4 shows the comparative experiment with and without applying the proposed phase-shift CPWM. Section 5 draws the conclusion of this article.

2. Double Fourier Integral Analysis Method of Pulse Width Modulation

According to [19], if the function of both $x(t)$ and $y(t)$ are periodic functions, the function $f(x,\ y)$ is the summation of sinusoidal harmonics. Assuming the periods of both the functions $x(t)$ and $y(t)$ are 2π, and $f(x,\ y)$ is represented by:

$$f(x,y) = \frac{A_{00}}{2} + \sum_{n=1}^{\infty}[A_{0n}\cos ny + B_{0n}\sin ny] + \sum_{m=1}^{\infty}[A_{m0}\cos mx + B_{m0}\sin mx] + \sum_{m=1}^{\infty}\sum_{n=-\infty}^{\infty}[A_{mn}\cos(mx+ny) + B_{mn}\sin(mx+ny)], \tag{1}$$

with:

$$A_{mn} = \frac{1}{2\pi^2}\int_{-\pi}^{\pi}\int_{-\pi}^{\pi} f(x,y)\cos(mx+ny)dxdy, \tag{2}$$

$$B_{mn} = \frac{1}{2\pi^2}\int_{-\pi}^{\pi}\int_{-\pi}^{\pi} f(x,y)\sin(mx+ny)dxdy. \tag{3}$$

where m and n are the positive integer numbers.

Sine-triangle modulation is one of the most commonly used CPWMs. Figure 2 describes the x-y plane of the sine-triangle modulation, which is represented by $f(x,y)$. The blue area represents the parts when the switch at the top of the converter leg is turned on, while the white area represents the parts when the switch at the bottom of the converter leg is turned on [28]. The x and y axes are considered as the carrier signal and the modulating signal time-varying angles respectively. $y(x)$ is described with the straight line in Figure 2a. The slope of $y(x)$ is the quotient between the frequencies of the modulating waveform and the carrier waveform. The intersection points between the function $y(x)$ and the changing boundaries of the function $f(x,y)$ are the switching time instants of the converter. The corresponding pulse width modulation (PWM) output voltage waveform (from $-\frac{V_{dc}}{2}$ to $\frac{V_{dc}}{2}$) is shown in Figure 2b.

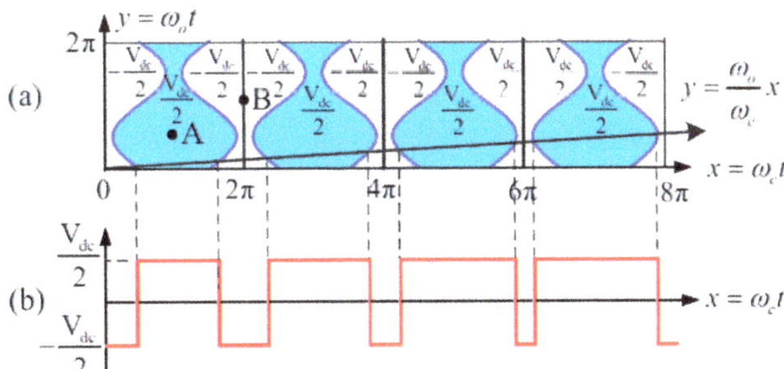

Figure 2. (**a**) The x-y plane for the sine-triangle modulation (**b**) The x-y plane for the corresponding pulse width modulation (PWM) output voltage [28].

The time varying modulating signal $V_m(t)$ can be presented as:

$$V_m(t) = Mcosy, \tag{4}$$

with respect to Equation (4), the intersection points in Figure 2a can be given as the following two cases. Case 1: for the function $f(x,y)$ switches from $-\frac{V_{dc}}{2}$ to $\frac{V_{dc}}{2}$, the switching instants happens at:

$$x = 2\pi p + \frac{\pi}{2}(1 + Mcosy),\ p = 0,\ 1,\ 2,\ 3,\ \ldots \tag{5}$$

Case 2: for the function $f(x,y)$ converts from $\frac{V_{dc}}{2}$ to $-\frac{V_{dc}}{2}$, the switching time happens at:

$$x = 2\pi p - \frac{\pi}{2}(1 + Mcosy),\ p = 0,\ 1,\ 2,\ 3,\ \ldots \tag{6}$$

The switching time instants in Equations (5) and (6) are the upper and the lower integral limits in Equations (2) and (3), respectively. The function of $f(x,y) = \frac{V_{dc}}{2}$ between the upper and the lower limits. Substituting Equations (5) and (6) into Equations (2) and (3), A_{mn} and B_{mn} can be rewritten as:

$$A_{mn} = \frac{1}{2\pi^2}\int_{-\pi}^{\pi}\int_{-\frac{\pi}{2}(1+Mcosy)}^{\frac{\pi}{2}(1+Mcosy)} \frac{V_{dc}}{2}\cos(mx + ny)dxdy, \tag{7}$$

$$B_{mn} = \frac{1}{2\pi^2}\int_{-\pi}^{\pi}\int_{-\frac{\pi}{2}(1+Mcosy)}^{\frac{\pi}{2}(1+Mcosy)} \frac{V_{dc}}{2}\sin(mx + ny)dxdy. \tag{8}$$

Substituting Equations (7) and (8) to Equation (1), and replacing x and y with $\omega_c t + \theta_c$ and $\omega_o t + \theta_o$ respectively, the time-varying output voltage of terminal a with respect to the ground in Figure 2 can be represented by the function of $f(t)$, which can be expressed as:

$$\begin{gathered} f(t) = \underbrace{\frac{V_{dc}}{2}}_{DC\ offset} + \underbrace{\frac{V_{dc}}{2}Mcos(\omega_o t + \theta_o) +}_{fundamental\ component} \\ \underbrace{\frac{4V_{dc}}{\pi}\sum_{m=1}^{\infty}\frac{1}{m}J_0\left(m\frac{\pi}{2}M\right)\sin m\frac{\pi}{2}cos(m[\omega_c t + \theta_c]) +}_{carrier\ harmonics} \\ \underbrace{\frac{4V_{dc}}{\pi}\sum_{m=1}^{\infty}\sum_{n=-\infty}^{\infty}\frac{1}{m}J_n\left(m\frac{\pi}{2}M\right)\sin\left([m+n]\frac{\pi}{2}\right)\cos(m[\omega_c t + \theta_c] + n[\omega_o t + \theta_o].}_{sideband\ harmonics} \end{gathered} \tag{9}$$

By using the double Fourier integral method, the time-varying phase leg voltage of $f(t)$ can be expressed with the addition of the dc offset, the fundamental component and all of the sinusoidal harmonic components (carrier harmonics and sideband harmonics).

3. Modelling of Equivalent Current Harmonics in Multi Three-Phase Drives

According to Equation (9), assuming the modulating signal start angle as the reference ($\theta_o = 0$), the phase leg voltage (phase leg voltage is defined as the voltage drop between a, b, c and z, which is displayed in Figure 1) of the pth, $p \in \{1, \ldots, N\}$ three-phase subsystem $u_{a_p z}$ (phase a) is represented as:

$$u_{a_p z}(t) = \frac{V_{dc}}{2}M\cos\omega_o t + \sum_{m=1}^{\infty}\sum_{n=-\infty}^{\infty} A_{mn}\cos\left\{mx_p(t) + n\omega_o t\right\}, \tag{10}$$

with:

$$A_{mn} = \frac{2V_{dc}}{m\pi}J_n\left(m\frac{\pi}{2}M\right)\sin\left[(m+n)\frac{\pi}{2}\right], \tag{11}$$

$$x_p(t) = \omega_c t + \theta_{c,p}, \tag{12}$$

where $x_p(t)$ is the carrier phase angle in the pth three-phase subsystem and $\theta_{c,p}$ is the pth three-phase subsystem carrier phase start angle ($t = 0$). Referring to Equation (10), the equivalent phase voltage $u_{total}(t)$ generated by all the subsystems (shown in Figure 1) can be represented as:

$$u_{total}(t) = \frac{NV_{dc}}{2} M \cos \omega_o t + \sum_{p=1}^{N} \sum_{m=1}^{\infty} \sum_{n=-\infty}^{\infty} A_{mn} \cos\{mx_p(t) + n\omega_o t\}, \tag{13}$$

As can be seen from Equation (13), the total equivalent harmonic voltage will be decreased with different carrier angles in different three-phase systems. Applying $\theta_{c,p} = \dfrac{2\pi(p-1)}{N}$ to each three-phase system, the multi three-phase drive system will gain an effective switching frequency of Nf_c. With the proposed phase-shifted CPWM in each three-phase drive, the total equivalent phase voltage is expressed by Equation (14):

$$u_{total}(t) = \frac{NV_{dc}}{2} M \cos \omega_o t + \sum_{p=1}^{N} \sum_{m=1}^{\infty} \sum_{n=-\infty}^{\infty} A_{mn} \cos\{Nmx_p(t) + n\omega_o t\}, \tag{14}$$

Comparing Equations (13) and (14), the equivalent voltage harmonic components are significantly mitigated, while the remaining harmonics only exist around the multiples of Nf_c.

Figure 3 shows the Fast Fourier Transformation (FFT) spectra of total equivalent voltage applying the phase-shifted CPWM with normalized V_{dc} with three modulation indexes ($M = 0.9$, $M = 0.5$, and $M = 0.1$). In Figure 3, the total equivalent voltage is supplied to the machine with the same rated power for any number of split three-phase systems, which means the fundamental voltage components are the same for different values of N. Figure 3 presents the FFT spectra of the total equivalent voltage without the proposed CPWM (for any number of split three-phase systems) at the layer of $N = 1$. Conforming to Equation (13), the total equivalent voltage is combined with the fundamental component and the groups of harmonics around the multiples of switching frequency. Figure 3 displays the FFT spectra of total equivalent voltage with proposed CPWM at the layers of $N \geq 2$. Consistent with Equation (14), the equivalent voltage harmonic components are significantly mitigated, while the only remaining harmonics exist at Nf_c. The larger value of N will result in higher order remaining harmonics, and thus the effect of total equivalent voltage harmonics cancellation is more dominant with the increasing value of subsystem number N. Comparing Figure 3a–c, it can be observed that a different modulation index affects the harmonic amplitudes to a different extent. However, the same groups of the harmonic components will be eliminated despite the modulation indexes, and thus the proposed phase-shifted CPWM method can be applied into different working conditions. Taking the case $N = 4$ as an example, with phase-shifted CPWM, the equivalent voltage harmonics around f_c, $2f_c$, $3f_c$, $5f_c$, $6f_c$, $7f_c$, $9f_c$, $10f_c$ are cancelled out under all the modulation indexes.

In [24], it indicates that the relationship between voltage harmonics u_h and current harmonics i_h can be drawn as:

$$u_h = i_h \cdot Z(h), \tag{15}$$

where $Z(h)$ is the impedance of the hth harmonic component. According to Equations (13) and (14), the equivalent phase current $i_{total}(t)$ can be represented as:

$$i_{total}(t) = \frac{NV_{dc}M}{2z(0)} \cos \omega_o t + \frac{1}{z(h)} \sum_{p=1}^{N} \sum_{m=1}^{\infty} \sum_{n=-\infty}^{\infty} A_{mn} \cos\{mx_p(t) + n\omega_o t\}. \tag{16}$$

Applying the proposed phase-shifted CPWM in each three-phase system, the equivalent phase current can be represented as:

$$i_{total}(t) = \frac{NV_{dc}M}{2z(0)} \cos \omega_o t + \frac{1}{z(h)} \sum_{p=1}^{N} \sum_{m=1}^{\infty} \sum_{n=-\infty}^{\infty} A_{mn} \cos\{Nmx_p(t) + n\omega_o t\}. \tag{17}$$

Referring to Equations (14) and (17), it is concluded that applying the proposed phase-shifted CPWM will effectively reduce the equivalent voltage harmonics as well as their corresponding current harmonics.

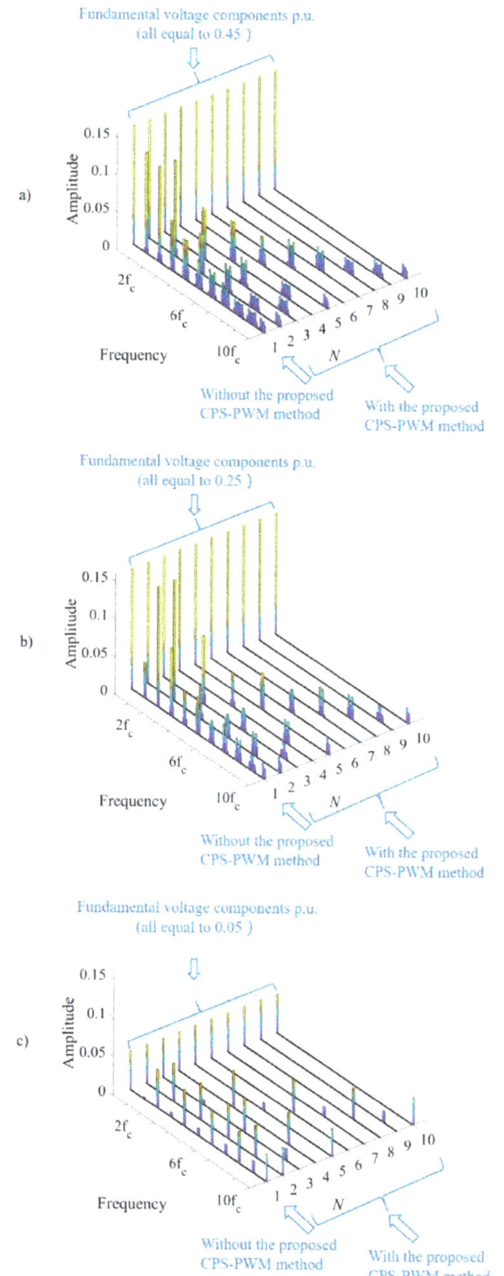

Figure 3. Multi three-phase drive FFT spectra with and without the proposed phase-shifted carrier-based pulse-width modulation (CPWM) under different M. (**a**) $M = 0.9$, (**b**) $M = 0.5$, (**c**) $M = 0.1$.

4. Experimental Results

A quadruple three-phase permanent magnet synchronous machine driven by quadruple independent three-phase modular inverters platform was set up in the laboratory, which is shown in Figure 4, to validate the proposed phase-shifted CPWM method analyzed in Section 3. The PLECS RT lab box is set as a real time controller to generate synchronized gate signals to drive the quadruple three-phase modular converters. The control strategy of this experimental test consists of two closed loops (i.e., the outer speed loop and inner current loop with proportional-integral (PI) controllers). The main experimental parameters are shown in Table 1. In order to better show the PWM harmonics, the fundamental components in Figures 5 and 6 are saturated.

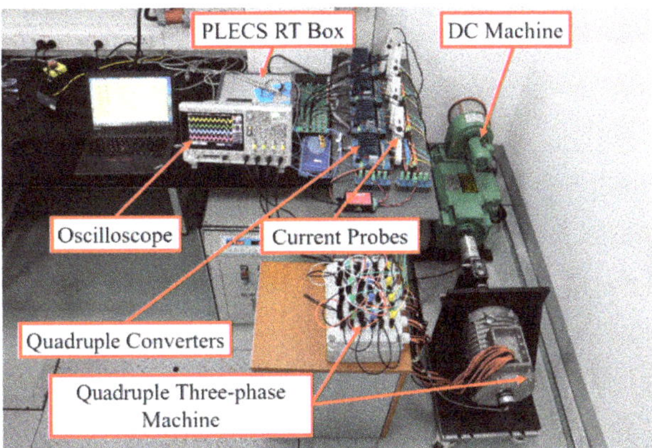

Figure 4. Quadruple three-phase drive system experimental setup.

Table 1. Experimental parameters.

Parameter	Value
Direct current (DC) link voltage (V_{dc})	40 V
Switching frequency (f_c)	2 kHz
Machine pole pair number	4
Load torque	3 Nm
Machine mechanical speed	200 rpm

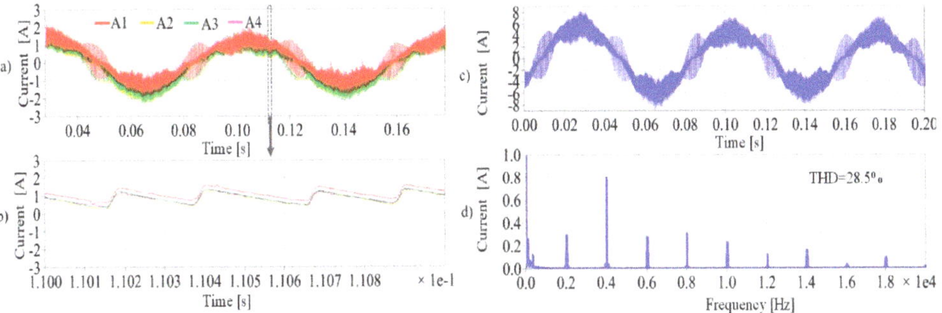

Figure 5. (**a**,**b**) Experimental result of phase currents without phase-shift CPWM. (**a**) Two fundamental period range. (**b**) The cursor range of Figure 5a,c,d. Equivalent phase current without phase-shift CPWM. (**c**) Time-domain current waveform. (**d**) FFT spectrum of Figure 5c.

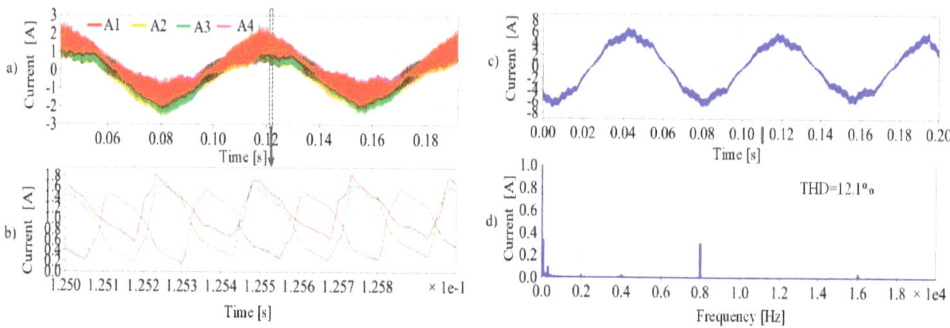

Figure 6. (**a**,**b**) Experimental result of phase currents with phase-shift CPWM. (**a**) Two fundamental period range. (**b**) The cursor range of Figure 6a. (**c**,**d**) Equivalent phase current with phase-shift CPWM. (**c**) Time-domain current waveform. (**d**) FFT spectrum of Figure 6c.

The phase currents experimental results without the proposed phase-shift CPWM are displayed in Figure 5. The carrier phase angles in each subsystem were set to be $\theta_{c1} = \theta_{c2} = \theta_{c3} = \theta_{c4} = 0$. Figure 5a shows the time-varying current waveforms (phase A) from all the four subsystems, while Figure 5b shows the cursor range of Figure 5a, which is the two-period range of the carrier signal. Figure 5a,b show that the phase currents are in phase in terms of both modulating the signal period range and carrier signal period. Figure 5c,d show the equivalent phase current and its FFT spectrum without phase-shifted CPWM.

The equivalent phase current waveform in Figure 5c is the same as the phase current waveforms in Figure 5a, and the current total harmonic distortion (THD) of both Figure 5a,c is 28.5%. This result matches with Equation (16) shown in Section 3, which indicates that the equivalent phase current magnitudes of both fundamental component and high-order current harmonics will be four times the phase currents' magnitudes in a quadruple three-phase drive. As shown in Figure 5d, the equivalent phase current harmonics exist at the multiples of the switching frequency (2 kHz), which exist around 2 kHz, 4 kHz, 6 kHz, 8 kHz, 10 kHz, 12 kHz, 14 kHz, 16 kHz, 18 kHz, 20 kHz. The harmonic FFT spectra are consistent with the analytical results shown in Figure 3 (the layer of $N = 1$).

Figure 6 describes the experimental results of phase currents applying the proposed phase-shift CPWM. According to the analytical models in Section 3, the carrier phase angles $\theta_{c1} = 0$, $\theta_{c2} = \frac{\pi}{2}$, $\theta_{c3} = \pi$ and $\theta_{c4} = \frac{3\pi}{2}$ are applied to the four subsystems accordingly. Similar to Figure 5, Figure 6a shows the four subsystem phase current waveforms, while Figure 6b presents the cursor range of Figure 6a. According to Figure 6b, it is can be seen that the phase currents in the carrier signal period range are effectively shifted with proposed phase-shifted CPWM, while they are synchronized in Figure 5b. Figure 6c,d show the equivalent phase currents waveform and its FFT spectrum applying the proposed phase-shift CPWM.

The current THD of Figure 6a,c are 36.1% and 12.1%, respectively. In terms of Figures 5a and 6a, it can be observed that the phase current THD is slightly increased with phase-shifted CPWM. This is mainly caused by the mutual coupling effect among machine stator windings, and the increase of phase current THD will result in a higher copper loss of the machine. Different machine stator layouts will result in different mutual coupling effect, which is worthwhile for further analysis in the future. However, this work is mainly aimed at the elimination of the equivalent phase currents to reduce the flux linkage harmonics, torque ripple, vibration and noise in the multi three-phase drives. Figure 6c shows that the equivalent current waveforms are significantly smoothed in respect with Figure 5c, and the current THD is reduced from 28.5% in Figure 5c to 12.1% in Figure 6c. Comparing Figure 5d,c, the majority of phase current harmonic components in Figure 6d are eliminated, except the harmonics around 8 kHz and 16 kHz, which are four times and eight times the switching frequency, respectively ($m = 1$ and $m = 2$). This result matches the analytical Equations (16) and (17) in Section 3 and the FFT spectra in Figure 3 (the layer of $N = 4$). Therefore, the experimental results have validated that the total equivalent phase currents can be effectively mitigated with the phase-shifted CPWM.

5. Conclusions

This work proposed a phase-shifted CPWM approach to eliminate the equivalent phase current harmonics in quadruple three-phase drives. The optimal carrier phase shift angles are proposed referring to the mathematical equivalent phase current models introduced in Section 3. There is more dominant phase current harmonics elimination with the increasing number of subsystems. The quadruple multi three-phase drive system with the proposed carrier phase angles of $\theta_{c1} = 0$, $\theta_{c2} = \frac{\pi}{2}$, $\theta_{c3} = \pi$ and $\theta_{c4} = \frac{3\pi}{2}$ is verified with experimental tests. The experimental results shows that, applying the proposed CPWM, the equivalent phase current harmonics components are effectively eliminated at the multiples of switching frequency, except the harmonic components around four times and eight times of the switching frequency. The equivalent phase current THD is reduced from 28.5% to 12.1% with the proposed phase-shifted CPWM. The main advantages of the proposed phase-shifted CPWM are listed as follows:

(1) The phase-shifted CPWM can be applied into arbitrary multi three-phase drives.
(2) The implementation of the proposed method is simple by changing the initial phase angles of carrier signals in the software.
(3) Compared with previous research works targeted at torque ripple reduction, this work aims to reduce the total equivalent phase current harmonics, which makes the analytical models much easier. Additionally, the elimination of total equivalent phase current harmonics will result in significant machine performance improvement in terms of torque, vibration and noise.

Author Contributions: Conceptualization, X.W. and C.G.; validation, X.W. and H.Y. resources, H.Z.; writing—original draft preparation, X.W.; writing—review and editing, G.B.; supervision, G.B.; all authors have read and agreed to the published version of the manuscript.

Funding: This research was funded in part by the Natural Science Foundation of Zhejiang Province under Grant LQ19E070002 and in part by the Ningbo Science and Technology Beauro under Grant 2018B10002 and 2018B10001

Conflicts of Interest: The authors declare no conflict of interest.

Nomenclature

V_{dc}	Converter dc-link voltage
M	Modulation index.
ω_o, ω_c	Frequencies of the modulating and the carrier signals respectively.
θ_o, θ_c	Initial phase angles of the modulating and the carrier signals respectively.
x, y	Time-varying angle of the carrier and the modulating signals respectively
J_n	Bessel function
f_c	Switching frequency
$Z(h)$	The hth harmonic impedance

References

1. Levi, E.; Bojoi, R.; Profumo, F.; Toliyat, H.A.; Williamson, S. Multiphase induction motor drives – a technology status review. *IET Electr. Power Appl.* **2007**, *1*, 485–516. [CrossRef]
2. Baneira, F.; Yepes, A.G.; Lopez, O.; Doval-Gandoy, J. Estimation Method of Stator Winding Temperature for Dual Three-Phase Machines Based on DC-Signal Injection. *IEEE Trans. Power Electron.* **2016**, *31*, 5141–5148. [CrossRef]
3. Abdel-Khalik, A.S.; Masoud, M.I.; Williams, B.W. Improved flux pattern with third harmonic injection for multiphase induction machines. *IEEE Trans. Power Electron.* **2012**, *27*, 1563–1578. [CrossRef]
4. Valente, G.; Formentini, A.; Papini, L.; Gerada, C.; Zanchetta, P. Performance Improvement of Bearingless Multisector PMSM With Optimal Robust Position Control. *IEEE Trans. Power Electron.* **2019**, *34*, 3575–3585. [CrossRef]

5. Dai, J.; Nam, S.W.; Pande, M.; Esmaeili, G. Medium-voltage current-source converter drives for marine propulsion system using a dual-winding synchronous machine. *IEEE Trans. Ind. Appl.* **2014**, *50*, 3971–3976. [CrossRef]
6. Bennett, J.W.; Atkinson, G.J.; Mecrow, B.C.; Atkinson, D.J. Fault-tolerant design considerations and control strategies for aerospace drives. *IEEE Trans. Ind. Electron.* **2012**, *59*, 2049–2058. [CrossRef]
7. Barrero, F.; Duran, M.J. Recent Advances in the Design, Modeling, and Control of Multiphase Machines—Part II. *IEEE Trans. Ind. Electron.* **2016**, *63*, 459–468. [CrossRef]
8. Lopez, O.; Alvarez, J.; Doval-Gandoy, J.; Freijedo, F.D. Multilevel Multiphase Space Vector PWM Algorithm With Switching State Redundancy. *IEEE Trans. Ind. Electron.* **2009**, *56*, 792–804. [CrossRef]
9. Levi, E.; Satiawan, I.N.W.; Bodo, N.; Jones, M. A space-vector modulation scheme for multilevel open-end winding five-phase drives. *IEEE Trans. Energy Convers.* **2012**, *27*, 1–10. [CrossRef]
10. Carrasco, G.; Silva, C.A. Space vector PWM method for five-phase two-level VSI with minimum harmonic injection in the overmodulation region. *IEEE Trans. Ind. Electron.* **2013**, *60*, 2042–2053. [CrossRef]
11. Ahmed, S.M.; Abu-Rub, H.; Salam, Z. Common-mode voltage elimination in a three-to-five-phase dual matrix converter feeding a five-phase open-end drive using space-vector modulation technique. *IEEE Trans. Ind. Electron.* **2015**, *62*, 6051–6063. [CrossRef]
12. Bodo, N.; Levi, E.; Jones, M. Investigation of carrier-based PWM techniques for a five-phase open-end winding drive topology. *IEEE Trans. Ind. Electron.* **2013**, *60*, 2054–2065. [CrossRef]
13. Yazdani, D.; Ali Khajehoddin, S.; Bakhshai, A.; Joós, G. Full Utilization of the Inverter in Split-Phase Drives by Means of a Dual Three-Phase Space Vector Classification Algorithm. *IEEE Trans. Ind. Electron.* **2009**, *56*, 120–129. [CrossRef]
14. Ren, Y.; Zhu, Z.Q. Enhancement of steady-state performance in direct-torque-controlled dual three-phase permanent-magnet synchronous machine drives with modified switching table. *IEEE Trans. Ind. Electron.* **2015**, *62*, 3338–3350. [CrossRef]
15. Patel, V.I.; Wang, J.; Wang, W.; Chen, X. Six-phase fractional-slot-per-pole-per-phase permanent-magnet machines with low space harmonics for electric vehicle application. *IEEE Trans. Ind. Appl.* **2014**, *50*, 2554–2563. [CrossRef]
16. Sala, G.; Girardini, P.; Mengoni, M.; Zarri, L.; Tani, A.; Serra, G. Comparison of fault tolerant control techniques for quadruple three-phase induction machines under open-circuit fault. In Proceedings of the 2017 IEEE 11th International Symposium on Diagnostics for Electrical Machines, Power Electronics and Drives (SDEMPED), Tinos, Greece, 29 August–1 September 2017; pp. 213–219. [CrossRef]
17. Calzo, G.L.; Zanchetta, P.; Gerada, C.; Gaeta, A.; Crescimbini, F. Converter topologies comparison for more electric aircrafts high speed Starter/Generator application. In Proceedings of the 2015 IEEE Energy Conversion Congress and Exposition (ECCE), Montreal, QC, Canada, 20–24 September 2015; pp. 3659–3666. [CrossRef]
18. Calzo, G.L.; Zanchetta, P.; Gerada, C.; Lidozzi, A.; Degano, M.; Crescimbini, F.; Solero, L. Performance evaluation of converter topologies for high speed Starter/Generator in aircraft applications. In Proceedings of the IECON 2014 40th Annual Conference of the IEEE Industrial Electronics Society, Dallas, TX, USA, 29 October–1 November 2014; pp. 1707–1712. [CrossRef]
19. Holmes, D.G.; Lipo, T.A. *Pulse Width Modulation for Power Converters: Principles and Practice*; Wiley: Hoboken, NJ, USA, 2003; ISBN 9780470546284.
20. Levi, E.; Dujic, D.; Jones, M.; Grandi, G. Analytical Determination of DC-Bus Utilization Limits in Multiphase VSI Supplied AC Drives. *IEEE Trans. Energy Convers.* **2008**, *23*, 433–443. [CrossRef]
21. Ryu, H.-M.; Kim, J.-H.; Sul, S.-K. Analysis of multiphase space vector pulse-width modulation based on multiple d-q spaces concept. *IEEE Trans. Power Electron.* **2005**, *20*, 1364–1371. [CrossRef]
22. Hammond, P.W. A new approach to enhance power quality for medium voltage AC drives. *IEEE Trans. Ind. Appl.* **1997**, *33*, 202–208. [CrossRef]
23. Fan, B.; Wang, K.; Zheng, Z.; Xu, L.; Li, Y. A new modulation method for a five-level hybrid-clamped inverter with reduced flying capacitor size. In Proceedings of the 2017 IEEE Energy Conversion Congress and Exposition (ECCE), Cincinnati, OH, USA, 1–5 October 2017; pp. 5262–5266. [CrossRef]
24. Wang, X.; Sala, G.; Zhang, H.; Gu, C.; Buticchi, G.; Formentini, A.; Gerada, C.; Wheeler, P. Torque Ripple Reduction in Sectored Multi Three-Phase Machines Based on PWM Carrier Phase Shift. *IEEE Trans. Ind. Electron.* **2020**, *67*, 4315–4325. [CrossRef]

25. Wang, X.; Yan, H.; Sala, G.; Buticchi, G.; Gu, C.; Zhao, W.; Xu, L.; Zhang, H. Selective Torque Harmonic Elimination for Dual Three-phase PMSMs Based on PWM Carrier Phase Shift. *IEEE Trans. Power Electron.* **2020**, 1. [CrossRef]
26. Valente, G.; Papini, L.; Formentini, A.; Gerada, C. Radial Force Control of Multi-Sector Permanent Magnet Machines for Vibration Suppression. *IEEE Trans. Ind. Electron.* **2018**, *65*, 5395–5405. [CrossRef]
27. Pan, Z.; Bkayrat, R.A. Modular motor/converter system topology with redundancy for high-speed, high-power motor applications. *IEEE Trans. Power Electron.* **2010**, *25*, 408–416. [CrossRef]
28. Liu, Z.; Zheng, Z.; Sudhoff, S.D.; Gu, C.; Li, Y. Reduction of Common-Mode Voltage in Multiphase Two-Level Inverters Using SPWM with Phase-Shifted Carriers. *IEEE Trans. Power Electron.* **2016**, *31*, 6631–6645. [CrossRef]

MDPI
St. Alban-Anlage 66
4052 Basel
Switzerland
Tel. +41 61 683 77 34
Fax +41 61 302 89 18
www.mdpi.com

Energies Editorial Office
E-mail: energies@mdpi.com
www.mdpi.com/journal/energies

www.ingramcontent.com/pod-product-compliance
Lightning Source LLC
LaVergne TN
LVHW070615170726
843515LV00004B/80

9783036515021